Globalization and Human Rights

IUS GENTIUM

COMPARATIVE PERSPECTIVES ON LAW AND JUSTICE

VOLUME 13

For further volumes:
http://www.springer.com/series/7888

Jesús Ballesteros
Encarnación Fernández Ruiz-Gálvez
Pedro Talavera
Editors

Globalization and Human Rights

Challenges and Answers from a European Perspective

Springer

Editors
Jesús Ballesteros
Philosophy of Law
University of Valencia
Campus Tarongers
Edificio Departamental Central
Avda. de Los Naranjos S/N
Valencia, Spain

Encarnación Fernández Ruiz-Gálvez
Philosophy of Law
University of Valencia
Valencia, Spain

Pedro Talavera
Philosophy of Law
University of Valencia
Valencia, Spain

ISBN 978-94-007-4019-8 ISBN 978-94-007-4020-4 (eBook)
DOI 10.1007/978-94-007-4020-4
Springer Dordrecht Heidelberg New York London

Library of Congress Control Number: 2012935856

Printed on acid-free paper

Springer is part of Springer Science+Business Media (www.springer.com)

Preface

The book is the result of a research project by the Spanish Ministry of Science and Education (2007–2010), as well as a Prometeo project for groups of excellence by the Generalitat Valenciana on "Human rights, sustainability and peace" (2010–2013).

The different chapters of the book have been in turn presented and discussed within the framework of two international conferences. The first called "Globalisation and international justice" was held at the UIMP (Valencia) in July 2007. Along with the authors of the book, researchers from the universities of Coruña, Padua, "Tor Vergata" Rome, and the International Institute of Human Rights in Strasbourg took part.

The second of these international seminars was held in Javea Forum, in November 2007. The theme was Global Justice and, along with seven of the book's authors, renowned jurists (university professors and magistrates) from Spain and North Africa: Morocco and Tunisia, took part. The seminar was aimed at young jurists from the Euro-Mediterranean area (one hundred young jurists from Spain, Italy, Israel, Jordan, Libya, Morocco and Tunisia).

Some of the abstracts were also discussed at other universities: in Coruña (Spain), at the "International Seminar on the Philosophy of Law: Globalisation, Law, Conflict" (March, 2010), organised by Prof. Pedro Serna, and at Teramo (Italy) in November 2010 on "The sign of Cain. Contemporary dimensions of bellic violence" organised by Prof. Guido Saraceni.

Likewise, the content of some chapters are speeches which have been discussed within the framework of the "International Seminar: What future after the economic recession and the Arab Spring?" (May, 2011) organised by the Master in Human Rights, Peace and Sustainable Development Course, Department of the Philosophy of Law, Morals and Politics of the University of Valencia, in which researchers from PRIO, IECAH and the universities of Navarra and Luiss Guido Carli, Rome, took part along with the authors of the book.

The publication of this book would not have been possible without the support of Professor Aniceto Masferrer, to whom we wish to express our gratitude.

Introduction

The book analyses some of the current threats to human rights, as well as to peace and development, posing them in relation to globalisation, and underlines their avoidable nature, for which a change of paradigm is becoming more and more necessary.

Within the complexity of the globalisation phenomenon, both positive and negative aspects may be distinguished. Among the former, the advances brought about by ICTs as regards facilitating worldwide communication are noteworthy. In this sense, the role played by social networks is significant in the recent protests and demonstrations in Arab countries which have meant hope for the recovery of the liberty and dignity of their people. On the other hand, the way in which globalization is actually coming about seems to be intimately connected to economism. This means a perversion of the economy into chrematistics and has made the predomination acquired by capital over other production factors, such as human nature and human work, possible.

This predomination of finances has had, and continues to have, diverse worrying aspects, and implies risks for human rights in as much as it hides what is real, due to the precedence of auto-referential speculation – which also relies on opaque financial instruments – over the true economy (production of goods and services) and over politics (such as cooperation for the common good).

The current economic recession is precisely due to this exorbitant growth of finances and also to the irresponsibility that permanent growth means for debt, both on behalf of public institutions and companies and families, due to the covering up of debt, under the appearance of credit. In the first case, the debt came from unsustainable military expenses stretching from the Vietnam war up to the invasions of Afghanistan and Iraq; in the second situation, to consumerism linked to a perception of the human being subjected to a desire of wealth and pleasure, fruit of the manipulation of marketing and a lack of personal reflection. This reality scam is connected with post-structural ideas, where the possibility of real references, with sense and transcendence, is denied, and where everything comes down to a game of signifiers (Chap. 1).

This hiding of reality has negative effects for the protection of human rights. From these, two paradigmatic situations are analysed in the first part of the book.

The first consists of the voluntary and culpable hiding from its respective contributory systems from which they recur to tax havens to practice the most genuine mode of exploitation: benefitting from social cooperation without participating in its responsibilities. It is, therefore, a false non-participation, a case of criminal self-marginalisation. Non-transparent financial systems, tax havens, must be regulated and controlled urgently, given that their very existence means the most radical negation of distributive justice (Chap. 2).

The other situation is the exclusion, against their will, of people with disabilities, who are the object of discrimination by those unaware of the differing abilities of others. In this case, it is the opposite of non-participation, for which society as a group is responsible. We have here a case of intolerable hetero-marginalisation, which impedes the recognition of the difference of disabled people as being compatible with equality of rights to make a truly inclusive society possible (Chap. 3).

The human rights affected in all of these situations are fundamentally the right to work, to a home, to participation and, at times – such as during the present recession, even the right to food. Economic globalisation has provoked the return of extreme poverty to northern hemisphere countries and has generated serious environmental consequences (destruction of non-renewable resources and the creation of non-recyclable residues).

The second part of the book deals with even more serious affronts to human rights: wars and the poverty associated with them that is the cause of the greatest number of avoidable deaths in the world. Here, what is directly threatened is the right to life and health, due to a lack of drinking water, food and medicines. These are threats due not to devaluation of politics and economies, but to their total disappearance in the face of violence. It is what is known as Failed States. Of them, there are two particularly significant examples: Afghanistan (Chap. 4) and Somalia (Chap. 5), both with wars that have lasted for the past three decades, and with devastating famines. Somalia is first in the *Foreign Policy* list of Failed States of 2010 and Afghanistan is number six, but both appear in first place as regards delegitimisation of the State.

Reflection on both experiences is useful for illuminating problems of great relevance to the current debate on human rights in the context of international relationships. The problem of global-local conflicts, with irrefutable global implications (think of the link between the current conflict in Afghanistan with the self-named "War on Terrorism", or the international repercussions of the actions of Somali pirates), but the devastating effects which develop and are suffered on a local scale, mainly by the civil population of the country (and this is something that is often ignored and/or hidden). The debate about the responsibility to protect, the tension between humanitarian intervention and national sovereignty, just war, along with the most efficient ways to fight under-development and poverty: fair trade and cooperation for development. The need to return to an atmosphere of politics in the most noble sense of the word to overcome violence by means of communication as an indispensable condition for the securing and consolidation of peace and to make effective respect for human rights possible.

This leads to the study of the current situation of the human right in continuity with the classic *ius gentium* and analyses its relationships with the *ius in bello*, – which declares all war illegal where there is death of non-warring people. Outlining just causes of war in the present day, taking into account the difficulties of having effective human rights in times of war, necessitates taking a fresh look at the problem of humanitarian rights. This is connected to the practice of human rights, that assists human beings in conditions of special vulnerability (Chap. 6).

The conclusions of these chapters highlight the avoidable nature of these threats, and therefore, the moral and legal urgency to fight against them. A change of paradigm is urged therefore, based on productive economy, compatible with ecological sustainability and with access to all resources, along with the precedence of politics as a capacity for carrying out activities together under the supremacy of the law.

In the third part, the European Union is presented as the geopolitcal area that is most respectful of human rights and as a possible model for the organisation of other areas, due to its awareness of the avoidability of social evils, in virtue of three fundamental motives.

In the first place, the analysis of the role of public opinion in the European Union highlights the importance of the same to a market society. This is due to the clearly non-economist project of the founding fathers, which holds the recognition of values such as human dignity, life and liberty as non-negotiable higher values, as such placed above economic interests and, therefore, susceptible only to mediation but not negotiation; the value of pluralism in society and the public role of religions within that, as areas that uncover aspects of the human being that escape governmental morals, such as suffering, fragility, vulnerability and solidarity itself. The pluralism that is at the root of the European Union should not be confused with relativism but should be interpreted as an opening for dialogue and intercultural exchange, which does not at all exclude the search for practical truth. On the other hand, the non-economism of the European Union is united with the importance of politics and States. "Europe cannot be a Europe of the citizens if it is not first a Europe of the towns, and it cannot be a Europe of the towns if not as Europe of the States" (Chap. 7).

In the second place, Europe is characterised by its long-term defence of a social State of Law, which cannot be confused with financial capitalism. It is true that there have been hiccups in the process, such as with the Maastricht Treaty, but the Lisbon Treaty insists on a social market economy (art. 3/3) that is based on the precedence of the productive sector over the financial one and on maintaining economic and social rights (Chap. 8).

In third place, the European Union is presented to the world as a "peace project". From its birth, founded on the reconciliation of the main participants in the Second World War, through the Treaty of Rome and the Economic Community, the "*European project*" has always been motivated by the search for peace as a priority. And that great (economical and political) project stems from the conviction that peace does not come from *dissuasion* but from *cooperation*; it is not born from *imposition* but from the *perception* of real expectations of development. This idea has allowed Europe to overcome the narrow paradigm of "stronghold", to become an open, plural, stable and solid community. That *European project* consists of creating an ever more ample 'community of peace and democracy'; it intends to remodel the world, establishing the proposals to make the Kantian dream of "perpetual peace" a reality, not based on the balance of military powers (or the supremacy of one of them) but on universal recognition of personal dignity, on the universal guarantee of human rights and sustainable human development, that implies overcoming the current neoliberal model of globalisation (Chap. 9).

Valencia, Spain

Jesús Ballesteros
Encarnación Fernández Ruiz-Gálvez
Pedro Talavera

Contents

Contributors

Jesús Ballesteros is Professor of Philosophy of Law and Director of Doctoral Programme: *Human Rights, Peace and Sustainable Development* at University of Valencia, Spain

Francesco Biondo is Tenured Full-time Lecturer of Philosophy of Law at University of Palermo, Italy

Jorge Cardona Llorens is Professor of International Law at University of Valencia, Spain and Member of the Committee on Rights of Children, U.N. Geneva

Encarnación Fernández Ruiz-Gálvez is Associate Professor of Philosophy of Law, Director of Departament of Law, Moral and Political Philosophy and Director of Master *Human Rights, Peace and Sustainable Development* at University of Valencia, Spain

Ana-Paz Garibo-Peyró is Tenured Full-time Lecturer of Philosophy of Law at University of Valencia, Spain

Pedro Talavera is Associate Professor of Philosophy of Law at University of Valencia, Spain

Isabel Trujillo is Professor of Philosophy of Law at University of Palermo, Italy

Ernesto J. Vidal Gil is Professor of Philosophy of Law at University of Valencia, Spain

Francesco Viola is Professor of Philosophy of Law at University of Palermo, Italy and President of the Italian Association for Legal Philosophy

Part I
Human Rights: Soft Threats

Chapter 1
Globalisation: From Chrematistic Rest to Humanist Wakefulness

Jesús Ballesteros

Abstract Chapter 1 analyses the epistemological, anthropological and ethical aspects of globalisation and, as a result of the 2008 crisis, the urgency for a change of paradigm. The proposal involves overcoming the decadent post-modernity that characterises post-structuralism – which abandons itself to the principle of pleasure, thus stimulating debt and speculation – by means of a humanist post-modernity based upon the following principles: in the epistemological field, the recuperation of reality in the face of the game (including the stock markets) and, therefore, the subordination of finances to productive economy; in the anthropological sphere, the recognition of the Golden rule in the face of individualism and the value of loyalty when undertaking commitments as opposed to instantaneism; in the ethical field, the defence of the economy of the market, avoiding market society, which is to say avoiding the corruption of the human being and the plundering of nature; and last but not least, the understanding that the fundamental cause of the current crisis is the absolutisation of money, for the latter cannot overcome the ontological finitude of human beings. Only man's sense of the transcendent can challenge the problem of economical equity adequately, by situating the person above things and currencies.

The phenomenon of globalisation is characterised by greater communication between individuals and peoples. Two aspects can be distinguished from this phenomenon: one which is essentially technological, the new information technologies and, in particular, the internet. Although, like all technological advances, this aspect is ambivalent, it can be considered fundamentally positive thanks to its capacity to allow people to be in contact with each other, albeit virtually, and to empower them to be able to fight against dictatorships. The Arab revolution, which

J. Ballesteros (✉)
Professor of Philosophy of Law and Director of Doctoral Programme: Human Rights, Peace and Sustainable Development at University of Valencia, Spain
e-mail: Jesus.Ballesteros@uv.es

J. Ballesteros et al. (eds.), *Globalisation and Human Rights*, Ius Gentium: Comparative Perspectives on Law and Justice 13, DOI 10.1007/978-94-007-4020-4_1,

is bringing liberty back to its peoples, would have been impossible without these technological means.

The other aspect is essentially economic and related to the total hegemony of capital; of the financial markets. This involves its freedom of movement and deregulation in a far more significant way than the other factors of production such as non-human nature and human labour. Capital moves freely beyond borders in a way which does not occur with people or with goods.

This article will focus exclusively on this second aspect, analysing in particular its cultural assumptions.

1.1 Economic Globalisation and Decadent Postmodernity

It is widely agreed that economic globalisation started in 1975 at the Rambouillet Conference[1] with the establishment of the free circulation of capital. This meant that the principle established at the Bretton Woods Conference, prohibiting the circulation of capital in order to avoid speculation, was abandoned. There is also considerable agreement on the fact that this deregulation of the economy in opposition to the agreements of Bretton Woods had been started 4 years earlier by the American President Richard Nixon on 15 August 1971, when he abandoned the principle of monetary stability which established the fixed convertibility of the dollar with gold. The process of economic globalisation was completed in 1995 with the creation of the WTO which would liberalise the trade of goods. Today, it is also commonplace to think that the still ongoing economic crisis of 2007 represents something for the deregulation that underpins globalisation which is akin to what the fall of the Berlin Wall meant for communism.

The hypothesis presented here is that in the development of this economic globalisation which is now in crisis, it is not enough to take into account the economist assumptions of neoliberalism, as is usually the case, but also – that which is less frequently studied – the nihilistic thinking of decadent postmodernity,[2] given that this is also located at the very basis of neoliberal proposals, as we will go on to demonstrate.

From this perspective, it also important to add to the date of 15 August 1971 the symbolic date of May 1968 in France. Both dates are related, albeit in different ways, to the Vietnam War.

---

[1]Chaired by Reagan, Thatcher and Pompidou.

[2]On this subject I refer to chapter three of my book *Postmodernity: Decadence or Resistance* (Pamplona: Universidad de Navarra, 1992), which deals with poststructuralism and decadent postmodernity.

1.1.1 May 1968 and Poststructuralism: The Loss of Referents and Deregulation

May 1968 in France carried various messages. Some of these are worthy of consideration, such as the rejection of war and the demand for greater democracy. But unfortunately, it was not these aspects which triumphed, but rather those which were attached to poststructuralism as a continuation of Freudian leftism and the anti-psychiatry of Reich and Marcuse. That is, the primacy of the pleasure principle over the principle of reality, of desire over reason, of the unconscious over consciousness, of the instant over the lasting, and against all commitment, all norms and all limits: "banning is banned". Poststructuralism, as both the mentor and the heir of May 1968 in France, has become one of the most hegemonic cultural elements of the last 40 years due to the presence and support it has been given in the mass-media. Poststructuralism has achieved such hegemony as a result of the dissemination and trivialisation of Nietzsche, who has been reduced to a defender of the instantaneous and playfulness, forgetting his ideas of *amor fati* and the eternal return. F. Nietzsche radically rejected guilt as much as possible, associating it with the reality of debt.[3] Despite its manifest anti-economicism,[4] the nihilistic influence of Nietzsche's thinking in this sense has contributed towards generalising debt in today's society.

The poststructuralist ingredients of economic Globalisation implied and continue to imply a negation of reality at an epistemological level, in line with the idea that there is nothing outside the text. "Il n´y a rien hors du texte".[5] Everything is reduced to interpretation or representation. Poststructuralism seems to respond perfectly to the decadence prophesied by T.S. Eliot in the following splendid verses: Where is the wisdom we lost in knowledge?, Where is the knowledge we lost in information?[6]

In effect, poststructuralism denies language any reference to a reality that would make knowledge possible. Moreover, it also denies language the possibility of finding the meaning of life, which is the basis of wisdom.[7] Therefore, this reduction of the real to mere information contributes to bringing about a progressive estrangement of reality (the real economy) and its replacement by the speculative game of the stock market, in other words, in the financialisation of the economy.

---

[3]*On the Genealogy of Morals*, book II, aph. 4. As is well known, *Schuld*, in German, designates debt and guilt together.

[4]"Superior men: leave the market-place" *Thus spoke Zarathustra*, ch. "On the Higher Man".

[5]J. Derrida, *De la grammatologie* (Paris: Minuit, 1967), p. 227. As argued by A. Llano, *Humanismo civico* (Barcelona: Ariel, 1999), p. 143, "We live in a culture that considers the whole of reality as representation and which revels in this "discovery" as if this entailed liberation from the harshness of life".

[6]Choruses from the Rock.

[7]On reference and meaning, see G. Frege, "Uber Sinn und Bedeutung," *Zeitschrift für philosophie und philosophische Kritik* NF 100 (1892): 25–50. As argued by A. Llano, *Metaphysics and Language* (New York: G. Olms, 2005), ch. I, ap. 3. Frege anticipates the most creative work of twentieth century philosophy, in particular Brentano's and Husserl's notion of intentionality.

At an anthropological level, poststructuralism means the negation of the subject as an existential and ethical unit and therefore of any responsibility before the other and the future. In the same way that there is nothing outside the text, neither is there anything outside unconscious and immediate desire. In contrast to Freud's beliefs, it is now the Id that has to triumph over reason. This instantaneism means the suppression of all moral limits, including the golden rule. There is no requirement to treat the other as oneself because, now that we are disparate, there is no temporal continuity of the I in time and, therefore, we are busy enough trying to look after our different Is.[8]

This poststructural influence has destroyed the Calvinistic roots of capitalism and turned it into ludocapitalism; a casino capitalism. Calvinist capitalism could be defined, as in R. Niebuhr's words,[9] as that of the moral man, dominated by the idea of duty in his private, family life (matrimonial faithfulness and responsibility to one's children) and in work (sobriety, economising, moderation) who finds himself installed in immoral society, ruled by the unbridled pursuit of profit.

Now, by contrast, what is dominant alongside instantaneism is playfulness, the taste for *vivere pericolosamente*[10] and, consequently, increasing irresponsibility. The mentality of the *homo ludens* displaces labour, effort and economising, and replaces these with risk as an element for achieving quick economic benefits and, at the same time, and as suggested by Huizinga,[11] means the primacy of the appearance, of the "as if", indifference and independence with regards to morals. The instantaneity of financial capitalism is not just based on the pursuit of profit (an aspect criticised by Keynes and Orlean[12]) but also on the sale of sex and privacy.[13]

---

[8]Ballesteros, J. *Postmodernity*, cit. Ch. 7.

[9]*Moral Man and Immoral Society: A Study in Ethics and Politics* (New York: Charles Schreibers's Sons, 1947).

[10]The *gefahrlich leben* exposed by F. Nietzsche in *Die fröhliche Wissenschaft*, aph. 208, as is well-known, turns into fascism and, in a market sense, infects the world from the 1970s. In *Zarathustra* he wrote: "Why are you interested in lost bets? Have faith in future ones". This playful sense of life, which links indebtedness with speculation, culminates today in the brokers who are creators of derived products such as Fabrice Tourre, Vice President of Goldman Sachs, The artifice of the ABACUS. 2007-ACI.

[11]*Homo ludens. Von Ursprung der Kultur mi Spiel* (Hamburg: Rowohlt, 1958).

[12]*Del´euphorie a la panique: Penser la crise financiere* (Paris: Editions E.N.S, 2009).

[13]In his book, *The Superstition of Divorce,* Chesterton warned that capitalism would destroy the family. On their part, R. Perpiña Grau, "La crisis de la economía liberal. De ethos económico al de la seguridad," in *De lo liberal y de los pueblos* (Madrid: Ed. Nacional, 1975), pp. 11–101 and J. Velarde, *El libertino y el espiritu del capitalismo* (Madrid: Pirámide, 1981), pointed out that the best way of uncovering the thesis of the ultra-liberals was to open the pages of pornographic magazines such as *Penthouse* (p. 13, note 14). See also the interview with M. Friedman carried out by Geoffrey Norman in *Playboy*, in FEB. 1973. Monetarism is responsible for the deregulation of the economy and for the financialisation of the economy. With Friedman, together with Leo Meladed, being, as we will see later, the founders of the International Monetary Market of Chicago in 1971 and, what is possibly more serious, of the commercialisation of society. Its influence overlaps with that of poststructuralism which negates all rules.

A remote theoretical precedent of this ludocapitalism can already be found in 1740 with the praise of extravagance and luxury carried out by Bernard de Mandeville.[14] In this way it was considered to be a priority to widen the rights that corresponded to simple adult desires without taking into account the consequences for third parties; those who are unable to defend themselves. This ludocapitalism would bring about market society or society of spectacle[15] in which intimacy is replaced by extimacy[16] and even more importantly, democracy is replaced by democrature.[17] While real rights to life and development become more vulnerable or disappear in the face of the volatility of the markets and commitment.

The death of the family as a reality based on the stable union of a man and a woman is declared, and even the prohibition of incest is now considered to be old hat. Similarly, the death of the company as a long-lasting project is also declared by subordinating it to the supreme interest of the shareholder, *shareholders value*, ignoring the affected groups (*stakeholders*).[18]

Taking into account poststructuralism when analysing the causes of the crisis enables an understanding which takes in all the problems which this crisis entails, which is to say both the phenomenon of indebtedness and speculation, in contrast to the ideological approaches that attempt to attend to different aspects of this question in a simplistic and unilateral way. Indebtedness and speculation are consequences of instantaneism itself which is encouraged by nihilistic premises. With the cult of speed and profit, with the negation of duration and commitment, the present jeopardises the future of human labour, the natural resources and money itself.

1.1.2 15 August 1971: The Trivialisation of Debt and the Stimulation of Speculation

On the 15 August 1971, Nixon decided to abandon the fixed exchange rate of 34 dollars per ounce of gold established at Bretton Woods. And on 18 December the Treaty of Washington is signed. The reason for this, according to

---

[14]In his writing of 1740, *The fable of the bees, or, private vices, public benefits* (1740), Penguin Books, 1970, Mandeville is also the author of an important book *A Modest Defence of Public Stews*, New York, Palgrave, MacMillan, 2006 and *Free Thoughts on Religion*, the *Church and National Happiness*, which is clearly antireligious.

[15]This is the title of the book by G. Debord, Buchet-Chastel, Paris, 1967.

[16]This expression comes from J. Lacan as a synonym for that which is banned from consciousness but which is now used in a completely different sense to refer to the trivialisation and venalisation of intimacy.

[17]In which reality is replaced by the "synthetic image of the media" and a "growing uniformisation" is produced which, through the "stimulation of laziness and comfort", leads to total immobilism. G. Mermet, *Democrature. Comment les médias transforment la démocratie* (Paris: Aubier, 1987), pp. 201 and 232ss.

[18]For a critique of this see P. H. Dembinski, *Finance servante ou finance trompeuse* (Paris: Desclée de Brouwer, 2008), ch. 3, aph. 4.

Daniel Bell,[19] was the enormous deficit in the balance of payments produced by the Vietnam War, as well as by the uncontrolled consumption of the North American people. The USA therefore decided unilaterally and arbitrarily not to meet the debt owed.[20] The connection between the lack of referents which characterises decadent postmodernity and the dollar's lack of basis as the benchmark currency is therefore manifested in this way. A convergence also emerges between the disappearance of the gold standard and the decline of the appreciation of the rule of gold, replaced by mere individual desire, the key element of poststructuralism.

Nixon's irresponsible decision had two immediate consequences:

(a) The negation of the current debt, which is to say of the convertibility of the dollar, gave the United States' Federal Reserve of private capital carte blanche to freely produce money. This creation of money was to be the oil that would ignite the current fire.[21]
(b) Given that speculation is based on obtaining profit based on the uncertainties of the future, currency fluctuations and changes in prices increase the possibility of the bet indefinitely.

Because of this, from 1971, the disappearance of the dollar's fixed exchange rate enables speculation on its future performance, thereby opening the way to an increase in "negotiation" and speculation on the participating currencies.[22] The USA, in effect, deregulates the stock market, producing a game without rules. This results in an increase in wholesale chrematism on the productive economy and politics. This means the inversion of the hierarchical order which accords with common sense and rationality as established by Aristotle.[23]

---

[19]In his book, *The Cultural Contradictions of Capitalism* (London: Heinemann Educational Books, 1975); a book that remains valid today and which can help to interpret the current crisis, p. 201 and subsequently. Requeijo also expresses similar ideas in his book *Odisea 2050. La economía mundial del siglo XXI*, Alianza, Madrid, 2009, whose title is rather inappropriate, but whose content is of great interest. And also Peter Drucker in 1965, in his anticipatory essay *Landmarkets of Tomorrow*. A *report on the New "Postmodern" World,* New Brunswick and London, 1999 2nd ed. pointed out the inherent risks of two North American inventions of 1776: paper money, which gives the State limitless power, and the popular army, which gives way to total war not limited to fair causes (p. 223). And he had deduced how war imposes terrible limitations on the economy, given the fact that the production of arms (especially aerial ones) requires an industry with full and exclusive dedication and which is very difficult to reconvert (p. 227).

[20]In their book, *Le capitalisme malade de sa monnaie*, Paris, F.X. De Guibert 2009, Edouard Husson and Norman Palma show that the effect of Nixon's measure, supported by Milton Friedman and criticised by James Tobin (p. 160), was to "briser le reel", given that "pour l'esprit de temps, le réel etait insupportable" (p. 165). At the same time, they highlight the fact that the measure was taken in response to the deficit created by the Vietnam War and that subsequent wars (Iraq, Afghanistan) have done nothing less than precipitate the current economic disaster (p. 10ss). See also Joseph Stiglitz and Linda Bilmes, *The Three Trillion Dollar War* (New York: Allen Lanes, 2008).

[21]R. Fernández Duran, *Capitalismo financiero y guerra permanente* (Madrid: Virus, 2003).

[22]E. Palazuelos, *La globalización financiera* (Madrid: Síntesis, 1989), 57–60 and 205–34.

[23]*Politics*, 1253a. On this, *Postmodernity*, ch. 2.

Nixon's decision to break with Bretton Woods was driven by Milton Friedman. The economist was quite clearly situated within the nihilistic dynamics of decadent postmodernity. In his work, he proposes[24] the decriminalisation of what he calls "victimless crimes" such as prostitution, drugs, obscenity, pornography, euthanasia, and abortion, given that if consent between adults exists, then no criminal injustice is committed,[25] and at the same time, high economic profit may be obtained from such activities.[26] He had already requested a break from Bretton Woods in 1953 in his article "A Proposal for Resolving the US Balance of Payments Problem" and he presented this directly to Nixon during his electoral campaign of 1968. As a result, 3 years after the presidential election, on the 13 November 1971, he guaranteed the Polish-Jewish financier, Leo Melamed, that Bretton Woods would never be re-established, thereby supporting the creation of the Currency Futures Market. In effect, the International Monetary Market was created in Chicago shortly afterwards, and was the result of the joint work of Leo Melamed and Milton Friedman,[27] transforming its Stock Market that traded in eggs, butter and tripe into the most radical futures market in history. In 1960, Friedman had already defended the positive and stabilising character of speculation[28] and in 1972 he explicitly defended speculation on the currency future markets.[29] Both of these characters' bibliographies reveal their enthusiasm for the speculative game in relation to currencies. Friedman wanted to bet against the pound sterling and Melamed had seen his father speculate with currency during his exile from Poland to Japan.

[24]Portrait: Milton Friedman, Challenge (May–June 1978), *Capitalism and Freedom* (Chicago University Press, 1962); *Free to choose* (New York, 1980).

[25]Robert Nozick thought the same. As we see later, *hedge funds* are based on the same premise.

[26]"Here is my final point. About drugs, about alcohol, about pornography and smoking and everything else. What business is it of yours what I do, read, buy, see, say, think, who I fuck, what I take into my body – as long as I do not harm another human being on this planet?"

[27]During the 1970s and 1980s, the economy was based on Friedman's theories. In fact he received the Nobel prize in 1976. His most fundamental idea was: "the business of business is business". By this he meant that the sole aim of a business is to create value and money for its shareholders. "*Free to choose*," Milton Friedman's timeless formulation, actually means the freedom of investors to profit on the upside of the market while coercing taxpayers into providing socialized protection from reckless investor risk-taking and greed. After years of inconclusive polemics on this issue ("free market" propagandists are fairly relentless, you know), it is refreshing, finally, to have a little clarity on this score. *Capitalist freedom*, it is now evident, means frenzied bursts of profit-driven recklessness punctuated by inevitable interludes of financial socialism. (Shhh . . . .) David Bollier, Paying for Wall Street's speculative greed, www//onthecommons.org/content.php?id=2228. On the Melamed- Friedman collusion, see C. Salas *La crisis explicada a sus victimas* (Barcelona: Alterna, 2009), Friedman's twilight was not seen until the end of the 1990s, when Greenspan managed to reduce unemployment to 4%, without this implying an increase in inflation in contrast to the thesis sustained by Friedman since 1968 that the existence of a natural rate of unemployment prevented inflation rises (NAIRU).

[28]In Defense of Destabilizing Speculation, reprinted as Ch. 13 in M. Friedman, *The Optimum Quantity of Money* (Chicago: Aldine, 1969).

[29]The need for futures Markets in Currencies in *The Future Marketing of Foreign Currencies*, Chicago, International Monetary Market of the Chicago Mercantile Exchange, 1971, pp. 6–12.

On creating the Currency Futures Market on 16 May 1972 in Chicago, Friedman wrote enthusiastically in *The Wall Street Journal*: "I am surprised to see that a gang of dice players, betting on tripe, are so reckless as to think that they could snatch business from some of the most sophisticated players in the world".

The key events that occurred in the new Chicago Market were:

1973, the stock market is opened for the purchase of stock options
1976, the futures market in gold begins
1978, speculation on the price of oil begins
1982, currency options begin

In 1983, the United Kingdom continues the gaming orgy with the *big bang* of the London Stock Exchange and the French left, led by Mitterand and Fabius, join the deregulatory movement. As Pierre Rimbert explains[30] "after the socialist and liberal turn of 1981, the stock exchange became an end in itself thanks to Laurent Fabius and his Minister for the Economy, Beregomov. Young men such as Jacques Attali, Jean Louis Bianco, Francois Stasse and Alain Minc, convinced Mitterand to deregulate the markets, arguing that the left had to do it because the right would not have done it". This exalting of the stock market as an end in itself, was formalised in the Law of 11/09/1985 which created the *French Financial Futures Market* (MATIF).[31]

The stock market was no longer in the service of produce, it stopped supporting companies that create employment and sought only the negotiability of everything according to its immediate profitability, universalising risk.

The marketing of the banking sector seduced people, making them believe that they had a right to satisfy their desires as long as they remained solvent, and it simultaneously convinced those who were insolvent that they could be solvent again by offloading their actual insolvency onto unknown third parties. As Bell argues,[32] "the trick of hire purchase – and of bank credits – was to avoid using the word debt and to highlight instead the word credit". As we can now see, this trick would be the key to financial engineering, by converting debt into negotiable credit.

---

[30] In his article, "La izquierda desregularizó los mercados. La paradoja francesa", *Le Monde diplomatique* May 2009, p. 7. The financialisation of the economy and its globalisation had already been seen in the "belle epoque" (1870–1914), together with the infamous Berlin Conference which distributed Africa among the great powers of the time and culminated with the First World War.

[31] The extent of its derivatives was as follows:

Holland EOE (European Options Exchange) 1978
United Kingdom LIFFE (London International Financial Futures Exchange) 1978
France MATIF (Marché a Terme International de France) 1985
Switzerland SOFFEX (Swiss Financial Futures Exchange) 1988
Germany DTB (Deutsche Terminbourse) 1990
Italy MIF (Mercato Italiano Futures) 1993

[32] *The Cultural Contradictions of Capitalism*, p. 76.

1.1.3 The Primacy of Derivatives. The Opacity of Finances and Information Asymmetry

From the 1970s onwards, an affinity between finances and poststructuralism is revealed in terms of their rejection of reality. This appears to be particularly clear with the primacy of derivative financial products over and above primary products. The idea of indefinite interpretation, of intertextuality,[33] which is characteristic of poststructuralists, connects perfectly with the primacy of the derivatives market, with indefinite speculation. Foucault writes that[34]: "If interpretation never ends, that means there is nothing to interpret. There is absolutely nothing fundamental to interpret given that, in essence, everything is already interpretation, every sign is not in itself the thing that offers itself up for interpretation, but rather the interpretation of other signs". We could replace the word interpretation with the word speculation and the word sign with the word financial product and we thereby find a description of what happens on the stock market and in banks: the self-referentiality of the speculative game in which reality (the company and the family) disappear.

In effect, from the end of the 1970s, the commercial bank, which collected deposits and invested them in non-risk short-term operations, broke the balance rule by creating *securities* in order to make its own credits immediately negotiable. In this way, the commercial banks wanted to imitate investment banks who always dealt with long-term risks, creating something on the lines of a bond that could be played on the stock market.[35]

In this way, an exponential increase was seen both in banks as well as on the Stock Exchange with the secondary market taking precedence over the primary market.

- A primary market is that in which trading occurs according to the number of shares ("pieces of a company"), private debt and public debt (the money that companies and governments borrow), raw materials, food and currency (the purchase or sale of foreign currencies).

---

[33] Intertextuality begins with Mallarme and is present in Barthes, Derrida and Kristeva.

[34] *Nietszche, Freud, Marx,* Ed. de Minuit, Paris, 1965.

[35] Minsky, -Wray, Securitization, *The Levy Economics Institute of Bard College*, 1987, p. 3. Liaw, *The Business of the Investment Bank* (New York: Willey, 1999). As Michael Lewis puts it in his book *Liar's Poker*, Penguin, 1990, the commercial bank managers were cattle ranchers and, on Wall Street, the cowboys were the best prepared. "The cowboys operated with public company bonds and the ranchers wanted to create something like bonds with securitisation".

- A secondary market is dominated by derivatives which consist in "wagers on the results of a future event".[36] The estimate of its value is supported by the price of other underlying assets. As already discussed, this began in Chicago with the foreign currency futures market. In all its forms, (1) futures, (2) options, (3) swaps, they offer an enormous possibility for getting rich out of high leverage, which is to say without any other obligation than that of having to pay in the future, without having to pay anything at the time of carrying out the operation.
  (a) Futures involve the purchase of the right to acquire something within X amount of time. The futures markets are responsible for increases in prices of raw materials and food products, with the subsequent effect this has on the increase in famine.[37]
  (b) Options also entail the possibility of acquiring something in X amount of time.[38] With these, a limit is fixed in relation to losses in value of the raw material while the benefits are unlimited. These are strictly speculative instruments, aimed at increasing the stock market value of shares. They are generally presented as an additional way of rewarding directors: they considerably increase the loyalty of executives who are key to the company and, if the stock market value of shares increases considerably, they can turn into a significant extra bonus without representing any apparent cost increase for the company. They provide the right to buy a number of shares at a predetermined price that can be executed after a certain number of years. The secret of their success is that once the beneficiary decides to execute their option – once the minimum period has elapsed and if the value exceeds the price established in the option – they buy and sell at the same time, or in other words they gain the difference. The ethical questionability of this presents itself when the beneficiaries are senior executives who may use their influence with regards to strategic decisions of the company in order to make personal profits. Stock options had a decisive role in the crisis of 2001, stemming form the Enron scandal.
  (c) *Swaps* and, in a very particular way, *credit default swaps* or CDS, are contracts in which *default risks* are swapped with an issuer of debt (*credit*). This involves making money by purchasing securities from companies or

[36]M. Roubini, *Economics Crisis. A Crash Course in the Future of Finance* (London: Allen Lane, 2010), ch. 8. this highlights how these provided cover against risk in the past but that now they have become something of a danger.

[37]López García and D. José Angel López López *Con la comida no se juega* (Madrid: Traficantes de sueños, 2003). As identified by E. Esposito, *Il futuro dei futuri, Il tempo del denaro nella finanza e nella societa* (Firenze: ETS, 2009), p. 210ss, the colonisation of the future has had disastrous consequences. See also, pp. 150s.

[38]L. Napoleoni, *La morsa. Le vere ragioni della crisi mondiale* (Milano: Chiarelettere, 2009), recalls that R. Merton and M. Scholes won the Nobel Prize in Economics thanks to their mathematical formula that calculated the value of the option, "the mother of all derivates". The following year, the investment fund managed by the Nobel prize winners, Long Term Capital Management, fell into ruin and had to be rescued by the Federal Reserve.

> states in danger.[39] These had a decisive influence on the crisis of 2007. In August 2007 over 45 trillion dollars were invested in CDSs, which is to say, more money than that invested in the stock market (around 22 trillion dollars), and with the GNP of the planet being approximately 64 trillion dollars. CDSs have been and remain the best example of how ludocapitalism renders companies and States impotent. Joseph Stiglitz[40] has shown why CDSs played such a harmful role in the current crisis. Firstly, "estimating the risk of a particular firm going bankrupt, however, is not like estimating life expectancy. AIG thought that it understood risk management. I did not". Secondly, "people weren't just buying insurance- they were gambling". "If one financial institution were to take out an insurance policy against Lehman Brothers dying, it would have an incentive to see Lehman's early demise. And there were ample weapons available to any player or group of players large enough to manipulate the market". Thirdly, the deeper contradiction: "Banks could not contemplate the possibility of an unpaid debt. However, they were buying and selling CDSs that were exclusively based on the high risk of non-payment".

The hegemony of derivatives goes hand in hand with the confusion between commercial banks that managed savings funds and investment banks that became legal in the USA in 1999 with the Gramm-Leach-Bliley Financial Services Modernization Act that revokes the Glass-Steagall Act of 1933.

The opacity of financial products can also be linked to poststructuralist jargon.[41] The financial market is one which epitomises most acutely Barthes's comment that: "The word is nothing more than a mask and all assemblies of it are simulacra".[42] What is particularly serious about finances is that, having reduced reality to pure information, the latter turns opaque once again and becomes inaccessible both for the treasury as well as for its own clients due to the asymmetry of information.[43] Current banks, which are both commercial banks and investment banks on behalf of companies who issue all kinds of securities, run into conflicts of interest given

---

[39] The principle takes us to the Maritime Capitalism of the Eighteenth Century when English people put wagers on the fact that certain ships would not return from the New World. And it turned out that people sank the ships in order to obtain the securities.

[40] *Freefall. Resource, Conflicts, Security and Global Justice* (London: Zed Books, 2007), ch. 6, p. 169ss. S. Zamagni states that "Even the most naïve economics student knows an economic law, the heritage of ancient wisdom, which says that the value of a complex financial product (think of the CDOs or CDSs) can never exceed the value of its weakest component – just as the strength of a chain is the strength of its weakest link". "The lesson and warning of a crisis foretold: political economy approach", *International Review of Economics* 56 (2009): 331.

[41] J. J. Goux, "Frivolité de la valeur. Essai sur l´imaginaire du capitalisme", "Values and Speculations: The Stock Exchange Paradigm," *Cultural Values* 1, no. 2 (1997): 159–77.

[42] *Le plaisir du texte* (Paris: Seuil, 1973), p. 129.

[43] On asymmetric information in finances and its consequences for human rights, see J. Tanega, "Investor's Rights to Disclosure of Complex Financial Instruments: A Risk Symmetric Analysis," in *Rights in Context. Law and Justice in Late Modern Society* (Surrey: Ashsgate, 2010), p. 301–16.

that they act both in the interests of the seller (the company issuing the securities) and the buyer of securities (a pensioner, for example) which implies playing with people.[44]

The clients' powerlessness increases with the opacity of the financial products and with the shadow banking which is basically constituted by *hedge funds*[45] and *private equity*. All these are off-balance sheet entities (O.B.S.E.)[46] that operate more than in any other way through OTC (*Over the Counter*)[47] transactions, or in totally deregulated markets outside the Stock Exchange, or through email or *e-commerce*. Their way of functioning is characterised by the instantaneism of the operations, by the short-term wagers that resemble sado-masochistic adventures: closely linked to leverage, they bet with securities that they do not possess and that imply 20 times their capital base in order to get rich when they fall.[48]

Finances are the perfect realisation of the fragmentation of the real and the mask which are characteristic features of decadent postmodernity.[49] Hence, in 1980, Sydney Homer of Salomon Brothers separated dividends from bonds in order to sell them both separately. Securitisation soon extended to the gigantic mortgage market, separating the payment of interest and capital in order to sell them as separate bonds. Everything becomes a source of open speculation without the need for money to buy securities.

Also contributing to the clients' powerlessness was the relativistic cynicism of the rating agencies, who defended themselves from their false ratings of financial products, claiming that they were mere opinions.[50]

The current gambling of the stock market, which is clearly self-referential, bets on the future of everything: currencies and gold, on how much financial products will go up or down and, what is even more serious, raw materials and food and now

[44]In the crisis of 2007, Goldman Sachs sold junk bonds to their clients at the same time that they were buying CDSs from AIG waging on the fall of the bonds that they had sold. In the script of the documentary *Inside Job* (p. 40), Lloyd Blankfein, Chairman of Goldman Sachs, argues that in the context of market-making these practices do not imply any conflict of interests.

[45]Created by Alfred Winslow Jones in 1949 in order to eliminate the risks involved in long-term investments by the short-term sale of shares. Winslow wrote in the magazine *Fortune*, which Daniel Bell considered "as the essence of the new hedonist capitalism", *Cultural Contradictions*, cit. p. 76.

[46]Of these kinds of funds, those of George Soros, John Paulson, Paul Tudor Jones (Tudor Investment) and Louis Moore Bacon (Moore Capital Management) stand out.

[47]For Jazmín Birnbreier, in her thesis "Análisis y evolución de la crisis financiera global de 2007/08" (thesis, ICADE, Madrid, 2009), p. 49, this is one of the dangerous characteristics of CDSs, alongside the fact that reserve capital is not required from the issuer on the grounds that it is not an actual security but a swap.

[48]Indebtedness and speculation are most widely expressed in FOREX (the Foreign Exchange), the electronic currency market with a daily volume of 1.5 trillion dollars. "More than 100,000 million against the Euro", *El economista*, 14/02/10.

[49]M. Roche insists on the opacity of Goldman Sachs, *La Banque. Comment Goldman Sachs dirige le monde* (Paris: Albin Mitchel, 2010), p. 275.

[50]*Inside Job*, script p. 43.

life assurance and junk bonds under a different name.[51] What is most alarming today is the increase in prices of raw food products due to the monopolisation of fertile soil, especially on the African continent. Entire crops are acquired for periods of 50 and more years by China and Saudi Arabia which is upsetting the international rates of basic products.[52] The 16.09.09 United Nations' World Food Programme warned that the number of people suffering from starvation had reached its historic peak by exceeding a billion. On the other hand, it is important to take into account that with less than 1% of that which governments have injected into the financial systems in order to pay for the gambling excesses of their respective countries, the problem of famine could have been resolved.[53] Altogether, volatility constitutes a violation against human rights as it increases vulnerability, poverty and marginalization.[54] And this is why it is a matter of urgency to forbid speculation in raw materials, food and energy products through public mechanisms to stabilise prices.

The lack of responsibility in the face of the problem of debt on the part of the USA[55] and the peripheral states of the European Union[56] has turned the current system into something of a "chimera",[57] revealing the monetary crisis to be another aspect of the crisis.[58]

---

[51]R. Pascual, "El ocaso de Wall Street", *Capital* 12, 10.09, pp. 36s. H. Kaufman in his book *The Road to Financial Reformation. Warnings consequences, Reforms* (John Wiley and Sons, New Jersey, 2009), p. 33s and 127s., criticising the thesis that there are too many large banks to go bankrupt and asking for the dismantling of the large financial bodies. According to data from El Economista on 25 June 2011, 85% of those who negotiate in wheat do not intend to make bread.

[52]As a denunciation of the International Food Policy Research Institute of Washington. A similar position is defended by economists such as M. Richardson, *Regulating Wall Street*, New York, Stern, 2010, as well as J. Stiglitz, *Free Fall.*

[53]The case of Somalia is particularly tragic which is analysed in Ch. 5 of this book. Somalia was forced by the World Bank and the IMF to change its self-sufficient agriculture for one of exportation. Drought and speculation with foods resulted in the monstrous famine of the summer of 2011.

[54]For further reading on this see F. Schumacher, *Small Is Beautiful, Economics as If People Mattered* (London: Blond and Briggs, 1973), p. 23ss. On the connection between volatility and vulnerability see J. I. Gonzalez, "La volatilidad acentúa la vulnerabilidad," *Innovar. Revista de Ciencias Administrativas y Sociales* (Universidad Nacional de Colombia, Bogotá), no. 21 (January–June 2003): pp. 25–30.

[55]Husson, E. Palma, N. Le capitalisme, malade de sa monnaie, cit.

[56]J. M. Daniel, *Le taureau face aux tigres. Entre les Etats-Unis et la Chine l'avenir est â l'Europe* (Paris: Pearson, 2007).

[57]It is no accident that the historian N. Ferguson described the symbiosis which existed in the last decade between North American debt and credit in China, and which appears to be reaching its culmination in China's reaction to the reduction of American debt by the agency *Standard and Poor's* in August 2011, as a Chimera. With regards to the risk affecting the Euro caused by the debt of the European periphery, the only way out is to achieve political and economic unity, which is to say, a United States of Europe, which was already proposed by V. Hugo, Ortega y Gasset, J. Habermas, and J. Ratzinger amongst others. On this I refer the reader to the third part of this book.

[58]For scholars such as D. Rodrik, *One Economics Many Recipes. Globalisation, Institutions and Economic Growth* (Princeton: Princeton University Press, 2007), pp. 199ss. R. B. Reich, *Supercapitalism. The Transformation of Business, Democracy and Everyday Life*, Alfred A. Knopf,

1.2 Resistant Postmodernity and the Economy of Permanence

1.2.1 The Return of Reality in the Face of the Financialisation of the Economy

There are authors who point out the relation between postmodernity and the economic crisis.[59] But most of these[60] are in no state to offer exit strategies as they are anchored to the poststructural approach or a decadent view of postmodernity, given that this professes a self-referential logic, one in which there is no reference to any reality or meaning. Following Baudrillard, they see the reduction of reality to simulation as inevitable, and far from enabling the possibility of change, they end up in a state of total hyperconformism (*hyper conformitè*).[61]

Because of this, in my book *Postmodernity, decadence or resistance,* I considered it necessary to oppose this kind of poststructural postmodernity with another model that reaffirms confidence in reason as the capacity to open oneself to the real, "*to the things themselves*", in the way opened up by phenomenology, and to try to understand the main challenges of today's reality, to seek out its meaning.[62] This recuperation of the referent or the real entails, in the case of the economy, the need to reject the self-referential and speculative nature of finances, the recognition of the humanist sense of the economy and its inseparability from the ethics of the duration and rejection of gigantism. We will now consider this in more detail.

The first philosophical demand of this resistant postmodernity is that of returning "to the things themselves", which means, as far as the economic sector is concerned, demanding a return to the real economy in the face of the primacy of finances.

(New York, 2007), p. 50s and 131s. They propose the creation of regional money reserves which would put an end to the supremacy of the dollar and ensure international financial stability.

[59]For example, E. G. Mc. Goun, "Hyperreal Finance," *Critical Perspectives in Accounting* 8 (1997): 97–122. See authors cited in footnotes 65 and 66. N. B. Mackintosh, "From Rationality to Hyperreality: Paradigm poker," *International Review of Financial Analysis* 12 (2003): 453–65. "A Linguistic Approach to Understanding Accounting: A "Wild card possibility", A. Hare, "A Shift Realized: The Banking Crisis as the First Postmodern Event," *Cultural Studies* 9, no. 2 (2009).

[60]One exception can be found in S. Sim's book, *The End of Modernity. What the Financial and Environmental Crisis is really Telling Us* (Edinburgh: Edinburgh University Press, 2010), p. 176ss.

[61]In the face of Baudrillard's sterile trivialities, the scientist Mandelbrot, whose concept of the fractal supposedly inspired the poststructuralist writer, contributed towards overcoming the economic crisis by showing the risk of the great oscillations in the financial market prices and how these should be corrected through regulation. *The (mis)Behavior of Markets. A Fractal View of Risk, Ruin and Reward* New York, Basic Books, 2004.

[62]*Postmodernity*, cit. Ch. 8.

In the twentieth century, there has been no shortage of economists critical of the phagocytosis of the economy by finances. A critic who was ahead of his time, and who discussed the reality of speculation that was already present in the *belle epoque*,[63] was Thortein Veblen who, in 1899, argued that the greatest social conflict was not, as Marx had believed, that between businessmen and their employees, but rather between the creators of wealth (business people and employees) and those who become rich without creating wealth (speculators as we would call them today), whose lifestyle revolved around gambling.[64] But this new capitalism would not, as we have argued, become hegemonic until the 1970s of the last century due to the influence of poststructuralism.

In 1936, much before the rise of speculation, J.M. Keynes presented the best critique of speculation in his main work *General Theory of Employment, Interest and Money*, by perceiving the ethical and anthropological degradation that it involves. "The effect of the investment markets has often led me to think that the purchase of an investment should be permanent and indissoluble, like marriage, except in cases of death or other serious circumstances".[65] Keynes warns here that the wrongfulness of speculation lies in its playful, instantaneous, short-term nature in a way which is comparable to sexual adventurism. Because of this, his proposals are based on a lucid comparison of the stock markets with casinos. "It is generally admitted that, out of public interest, casinos should be inaccessible and costly, and the reason why the *City* is less dangerous than *Wall Street* is precisely because of the more limited nature of the former". To this end, he considers it appropriate to impose high taxes on stock market transactions in order to limit their risks.

Fifty years later, in 1986, three eminent economists would highlight the risks of the growth in stock market speculation associated with investment banking.

---

[63]On the role of the city in this context, see P. J. Cain and A. G. Hopkins, "Gentlemanly Capitalism and British Expansion Overseas, II, New Imperialism, 1850–1945," *Economic History Review* XL (1987): pp. 3ss. The article begins with a quotation from W. Churchill, "I would see Finance less proud and Industry more content".

[64]*The Theory of Leisure Class. An Economic Study of Institutions* (New York: B. W. Huebsch, 1919). Thorstein Veblen's great achievement, both in this book and in his article "The instinct of Workmanship and the Irksomeness of Labor," *The American Journal of Sociology* 4 (1898–99): 187–201, lies in having deduced, in contrast to Marx, that the great contradiction was not produced between business people and workers but rather between those who live in peace, work, create wealth, and do not squander, and the parasites who prey, first through aggressive or warmongering exploits and later through speculation. His conceptions of "conspicuous waste" and "industrial sabotage" to describe finances are just as apt today in relation to the present crisis. On this, see John Atkins Hobson, *Veblen*, (London: Chapman, 1936). His book, *The Nature of Peace* is, by contrast, overly simplistic and anti-German.

[65]*General Theory of Employment, Interest and Money*, ch. 12, ap. 6. It is particularly significant that Keynes highlights the primacy of the spirit of matrimonial commitment as a key element and in contrast to the irresponsibility of the casino when discussing the health of the economy.

Peter Drucker[66] highlights the separation between finances and the real economy showing how the size of the former exceeds the size of the latter in a way which is alarming. "Every day the London Interbanking Market transfers 10 or 15 times the total amount of transnational transactions necessary to finance the exchanges between goods and services. Ninety percent or more of the financial transactions made by the transnational economy do not serve what economists consider to be an economic function: they serve purely financial functions". What Drucker calls the symbolic economy significantly dominates the real productive economy. Sixteen years later,[67] Peter Drucker states that when the commerce generated by the new financial services in the 1970s became the main activity, it was no longer commerce and turned into a "gamble", adding, "until the losses were so enormous that it was no longer possible to hide them, the speculator was a hero and he swam in money". And, he goes on: "The international money market does not produce anything. It is non-function money. It cannot generate any benefit because it does not serve any function (i.e.: the London Interbanking Market generates every day more movement of dollars than the whole world would need to finance all the economic transactions during 1 year)". A very important part of this virtual money, which has maximum volatility, is constituted by the foreign currencies market.[68]

Also frequently quote today is Hyman Minsky,[69] who argued that financial crises are related to recklessness when the bets are raised. When the economy is in a period of relative stability, everyone assumes that conditions will remain stable indefinitely. Gradually, they take on greater risks, with some requesting and other granting more and more loans, which, little by little, produces instability, since the instability of the economy is measured by the percentage of financing that fits into each one of the following three categories:

1. "financing with private insurance coverage", investors are able to repay their loans with the money which they earn.
2. "speculative financing", the physical money earned only pays off the interests of the loan and the main part has to be refinanced by new loans.
3. "Ponzi Financing" (which gets its name from the famous fraudster of the 20s, Charles Ponzi), the physical money generated is not enough to pay either the interests or the main credit and new loans are needed to meet all of a company's payment obligations. The only way out is to increase the value of the assets in

[66]In his article, "The Changed World Economy," *Foreign Affairs* 64–4, (Spring 1986): p. 768ss and subsequently in his book *The New Realities. In Government and Politics, in Economy and Business, in Society and World View* (New York: Harper & Row, 1989).

[67]*Managing in the Next Society* (Oxford: Butterworth and Heineman, 2002), pp. 93–5 and 106–8. This view is shared by the French Nobel prize-holder in Economics, Maurice Allais.

[68]According to data from the Basel Bank, in 2009, 1,000 trillion dollars of fictitious capital circulated in the world, while the world GNP was no more than 50 trillion dollars.

[69]*Stabilizing an Unstable Economy* (New York: MC.Graw Hill, 2008), pp. 230–45 and 352–61. (1st ed. Yale University Press, 1986).

order to obtain new loans. In the end, the phenomenon gives was to an economy which is heavily indebted, highly unstable and prone to crisis; an economy which is mainly based on "Ponzi financing" will be the most unstable kind and the kind which is most prone to crisis. In the years leading up to the crisis of 2008, in the global economy and particularly in the capital markets of the United States, the risks of "Ponzi financing" have increased at an alarming rate.[70]

Finally, Susan Strange[71] coined the term *Casino Capitalism*[72] to highlight the point that the primacy of finances over production turned the economy into a grand casino in which the few who participate determine the life of the entire population. In her book, Strange shows how the following components have emerged in casino capitalism since 1970.

(a) a pseudo monetarism due to the uncontrolled growth of public expenditure, especially in terms of arms and wars[73]
(b) a total deregulation of finances that allows for speculation with new products

In her book of 1998, *Mad money*,[74] this same author predicted the crash of 2008 based on the aggravation of the irrational and unpredictable nature of the financial system. But unfortunately her warning was not heeded.

Neither was John Paul II listened to when, in his speech on globalisation for World Peace Day on 1 January of the same year, he warned of the financialisation of the economy and the crisis of the social State. The following year[75] he criticised the fact that while there is a clear relation between the amount of labour carried out and the amount of goods produced in the economy of production, enormous sums of money can be obtained in the financial markets without the need for much labour at all. He also highlighted the demand for regulation of the financial sector on the basis that it is always those who are most weak who pay the highest price in times of crisis.

Since the crisis, the two eminent humanist economists, Dembinski[76] and Zamagni, have criticised the financialisation of the economy in a highly perceptive way.

---

[70] *Stabilizing,* cit. pp. 225–38. H. Minsky is an important critic of financial capitalism, or *capitalism management*, highlighting correctly that since the 1970s, finances have taken over business managers, *Finance and Stability. The Limits of Capitalism*, Milan, 1993. His accute analysis of the financialisation of the economy leads him, perhaps, to see the real economy in second place when he claims that "all economic crises are preceded by a financial crisis" (*Fundamental Reappraisal of the Discount Mechanism. Financial Instability Revisited: The Economics of Disaster*, 1970, p. 62).

[71] Professor in International Economics at the University of Warwick (+2000).

[72] As indicated in footnote 51, J. M. Keynes, in *The General Theory*, had already referred to Wall Street as a casino, pp. 145ss.

[73] S. Strange, *Casino Capitalism* (Manchester: Manchester University Press, 1986), p. 56.

[74] *When Markets Outgrow Government* (Ann Arbor: University of Michigan Press, 1988).

[75] From a speech addressed to the members of the Centessimus Annus Foundation on 11 September.

[76] *Finance servant ou finance trompeuse*, cit.

Stefano Zamagni[77] in his article on the crisis,[78] argues that it is not a temporary or regional problem but rather that it has a systemic character; it affects what we call globalisation in its entirety, and it is due to the disproportionate role that finances play within it. In other words, the key point is to defend the market economy by organising economic activity which is governed by the principle of the common good, according to which the good of each and every individual is vital, and to reject capitalism as a means of structuring society[79] that implies a general well-being that, in line with Bentham's utilitarianism, allows some to be excluded on the basis of the greatest good of the greatest number.

1.2.2 The Humanist Sense of the Economy: Against Market Society

Proceeding further, we will now highlight the humanist character of the economy and offer a critique of market society. The classical contributions of civil economy, as well as the contemporary approaches of Ordoliberalism and the thinking of F. Schumacher, will be of particular help in order to carry out this task.

What is referred to as civil economy, whose formulations can be traced back to the Franciscans' ideas of the fourteenth century, and much more specifically to those defended by Saint Bernardino of Siena (1380–1444), proposes a humanist conception of market economy founded on the following civic virtues.[80]

(a) Labour as something which is essential for the fulfilment of the individual, not just as a means of survival. What is important is to have a job and not to receive benefits, given that, as the Franciscans of the fourteenth century argued, "handouts help to survive but not to live. Because living means producing and aid (or benefits) do not produce".

---

[77]S. Zamagni and L. Bruni, *Economia civile. Efficienza, equità, felicità publica* (Bolonia: Mulino, 2004).

[78]"The Lesson and Warning of a Crisis Foretold: a Political Economy Approach," *International Review of Economics* 56 (2009): 331. "Since the phenomenon that we call globalisation started, finances have not only increased constantly their activity quota in the economic field, but they have contributed to modify both the cognitive positions of people and their system of values. Those who speak about financialisation of society refer to the latter aspect". The nucleus of this way of thinking, which subordinates everything to speculation, is that belief, as Keynes already anticipated, has been taken over by a fallacy. Zamagni's analysis highlights the pernicious global consequences of this process of the financialisation of the economy and society, that turns means into ends and ends into means, and stresses the urgent need to put a stop to the self-referential nature of finances.

[79]S. Zamagni and V. Zamagni, *La cooperazione* (Bolonia: Mulino, 2008), p. 15 and 21s.

[80]*Prediche Volgari sul Campo de Siena*, Milan, 189, cit. by L. Bruni and S. Zamagni, *Dizionario di economia civile* (Rome: Citta Nuova, 2009), pp. 107ss.

(b) The freedom of business, the capacity to assume risks responsibly and to be able to innovate, as well as to organise and coordinate. And the principle of subsidiarity.
(c) Development, leaving future generations more than that which has been inherited from past generations, which is to say intergenerational solidarity, equity and fraternity.

At the same time, civic economy firmly rejected a market society or capitalism which subordinated all to money, which ended up putting usury at the centre of everything, thereby strangling society instead of making it grow, and meaning that production was merely aimed at generating profit and not at seeking the common good. As we can see, despite the significant passage of time, these principles remain highly relevant today.

The central idea of civil economy, the distinction between an ethics-based market economy and economistic capitalism, was defended in the 1930s by the Freiburg School and Ordoliberalism.[81] Their main idea is the defence of the humanist goal of the economy and the critique of quantitativism, of the forgetting of the qualitative which is characteristic of economism. Röpke writes that[82]: "The cult of the standard of living is the social crisis of our times; the breakdown of the morals and spiritual foundations of society". Hence the rejection of the mathematisation of the economy that has led to the exclusion of the human being, to the forgetting of the freedom of human action and to a surrendering to determinism.[83] In line with this, Ordoliberalism highlighted the importance of qualitative aspects, fundamental aspects such as the understanding of the ecological crisis, as we will see later. One of the astute axioms of the School is their insistence on the primacy of Law, of the demand for regulation of the economy against the "spontaneous order" defended by neoliberals. On this aspect, Walter Eucken is particularly insistent.

[81]This is based on the social market economy of the "German miracle economy" of the 40s to the 70s. Its most significant representatives were Wilhelm Röpke and Walter Eucken as well as Ludwig Erhard.

[82]*Jenseits von Angebot und Nachfrag*, Bern, 1958, now, Verlagsanstal Handwerk, Düsseldorf, 2009, ch. 3, 2. In this same sense, Michael Lewitt warns that the economic crisis was a result of the central role played by *quants*, "quantitiative risk analysts", who are totally indifferent to the real economy and only focus on the increase in quick and unlimited profits. It is precisely because of the absence of quants from emerging economies that their emergence has been possible (The first thing we need to do is get rid of quants, *El economista*, 30.05.10. p. 17.) On his part, A.G. Haldane, "Rethinking the Financial Network" in *Financial Student Association*, Ámsterdam, April 2009, has underlined the interrelation between instability and the heterogeneity of financial agents, proposing a return to the Glass-Steagall Act which separated commercial banking from investment banking.

[83]Mathematised in this way, the economy turns into a Sad Science. This happens with the approaches of Walras and Pareto as well as with the utilitarians. In the face of this mathematisation and the false dichotomies it produces (between truth and reality, between necessities and profits) paradox reigns. Cfr. W. Eucken, *Grundlagen der Nationalökonomie* (Berlin: Springer, 1950) cit. passim, W. Röpke, *Die Gesselschaftskrisis der Gegenwart* (Erlenbach-Zurich: Eugen Rentsch, 1942).

"There can be no free economy without social virtues. The social market economy is a product of culture which corresponds to political democracy to the extent that it is highly difficult to construct".[84] Its existence means that human beings are not mere competitors, producers or negotiators, but simply human beings. "The market economy can only be sustained in a non-commercialised society".[85] Eucken highlights the need for an Economic Constitution in which "a sustainable economy befitting man" would be guaranteed by law and in which the responsibility of business people would be unlimited in order to ensure that their actions are prudent and do not cause any damage to third parties in cases of bankruptcy.[86] This very distinction between market economy or business economy and capitalism appears in this text of *Centessimus Annus:* "If by "capitalism" is meant an economic system which recognizes the fundamental and positive role of business, the market, private property and the resulting responsibility for the means of production, as well as free human creativity in the economic sector, then the answer is certainly in the affirmative, even though it would perhaps be more appropriate to speak of a "business economy", "market economy" or simply "free economy". But if by "capitalism" is meant a system in which freedom in the economic sector is not circumscribed within a strong juridical framework which places it at the service of human freedom in its totality, and which sees it as a particular aspect of that freedom, the core of which is ethical and religious, then the reply is certainly negative".

The primacy of Law, demands an end to the financial sector's lack of transparency. This would mean that all banking activities would have to be carried out on the balance sheet and, consequently, all *hedge funds* would also be controlled[87] as well as all financial instrument such as CDOs, CDSs, shortfalls without financial coverage against shares at risk such as in OTC derivative markets.[88] In other words, an attempt is needed to put an end to shadow banking, with financial paradises and

[84]*Jenseits von Angebot und Nachfrag*, cit. Ch. 3, 2.

[85]*Civitas Humana. A Human Order of Society* (Londong: William Hodge, 1949), p. 27. Another key point of interest of Ordoliberal thinking lies in their radical opposition to war and to military expenditure. They were the first to perceive the importance of the peace dividend for development. *International Order and Economic Integration* (Holland: Dordrecht, 1959).

[86]*Grundlagen der Nationalökonomie,cit.* pp. 52–54. This is precisely the opposite principle to the "too big to fall" idea defended by the Bush and Obama administrations with regards to the large financial groups. Ordoliberalism thereby coincides with the approaches of Hillaire Belloc and G.K. Chesterton, especially in his work of the noteworthy title *The Outline of Sanity*, as well as with the liberal interventionist ideas of A. Rüstow.

[87]M. Roubini, *Economics Crisis. A Crash Course in the Future of Finance* (London: Allen Lane, 2010), ch. 210 and 50, pp. 273 and 80.

[88]The ex-European Parliament speaker on banking regulations, and today Spanish Foreign Affairs and Cooperation J.M. Garcia Margallo, believes that the European Parliament must create a European Authority to supervise national and transnational banks, insurance companies and stock markets, and to control all hedge funds, including those of the USA. Borders should not impose any limits on supervisors given that these do not exist for hedge funds: if this regulation is not applied in its entirety, he concludes, we will end up back where we were before the Treaty of Rome.

Offshore operations.[89] In this sense, recent G20 summits have failed in relation to banking regulations due to the differences between Europe on the one hand and the USA and the emerging countries on the other. Alongside the USA and the United Kingdom, up to now emerging countries have been opposed to the taxation of banking, claiming that banks should only pay when they act irresponsibly, which is to say in advanced countries, while they assume that they have converted capital flow into production investments by sanctioning speculative movements (the so-called "flights of capital").[90]

1.2.3 The Critique of Collosalism

Capitalism and its extension into globalisation, is characterised by colossalism in the face of the market economy which is the organization of competition. This collosalism is the result of the scientific belief in the superiority of the visual over and above the oral, of the quantitative over the qualitative, of mathematics over human experience. The concentration of capital is linked to mental dispersion, to the absence of inner life and of critical spirit. As Röpke argues that Capitalism is opposed to the market economy because this excludes monopolies and oligopolies.[91]

The crisis of 2007 demonstrates the validity of the Ordoliberal assertions by showing the economies of scale, of the greater the better, to be obselete.[92] Collosalism had appeared in the 1970s with the multinational corporations who controlled 11% of worldwide production but represented 2/3 of world trade,[93] pursuing the best share performance at the same time as they froze salaries.[94] In the financial sphere, the principle of "Too big and interconnected to fall"[95] has led to the impunity of those responsible for the disaster, of the large investment banks, and at the same time has meant the transfer of debt from the banks to states, producing the fiscal crisis of the latter with the risk that several of them may fall into default. Thus,

---

[89]On this see Ch. 2 of this book.

[90]This was already signalled by the World Bank in their report of 1993, cit. by J. I. Gonzalez, "La volatilidad acentúa la vulnerabilidad," *Innovar. Revista de Ciencias Administrativas y Sociales* (Universidad Nacional de Colombia, Bogotá), no. 21 (January–June 2003): 15 and 21s.

[91]*Civitas Humana,* cit. introduction.

[92]Schumacher, F. *Small is beautiful*, 1/4.

[93]Dembinski, P.H. *Finance Servant*, cit. Ch. 6.

[94]P. Blond, *Red Tory: How Left and Right Have Broken Britain and How We Can Fix It* (London: Faber, 2010).

[95]*Economics Crisis. Economics Crisis. A Crash Course in the Future of Finance*, cit. ch. 210 and 50 pp. 210 and 50. A. Orlean insists that the freedom of the movement of capital is a main cause of the crisis, *Del´euphorie a la panique, cit.* p. 93 and 100, affirming that the pursuit of instant profit by investors is more serious than the opacity of the products.

today, the profits of the managers of large banks are multiplied while unemployment grows, since the small and medium businesses, which create 90% of employment, cannot get credit.

It is therefore crucial to reinstate the human dimension into business and into credit in order to re-establish competition and to reduce inequalities.[96]

1.2.4 Responsibility for Human Nature and Non-human Nature

The gigantism of globalisation seeks to fill the vacuum left by instantaneism. As already argued, "gigantism is a mask for volatility".[97] The shrinking of time, reduced to the instant along with the extension of space to the global sphere are additional features of the mathematisation and reification of reality, of the loss of the world of permanence and of humanity,[98] which are also inherent features of poststructuralism.

This instantaneism prevents any comprehension of the gravity of the current crisis, which is not only financial, fiscal or economical, but also ecological, driven by the destruction of non-renewable resources and the creation of non-recyclable waste. This crisis obeys the false belief in indefinite growth. The economy is a subsystem of the ecological system, all its income is derived from the environment and therefore indefinite growth is impossible. Up to the Industrial Revolution, humanity maintained itself with the annual energy produced by the sun; photosynthesis. Since the Industrial Revolution we have lived through an era of credit, using the resources accumulated over millions of years in the form of fossil fuels over the period of two centuries.[99]

Because of this, since the Stockholm Conference of 1972, there has been an insistence on sustainability, which means the recuperation of the qualitative dimension: the distinction between renewable/non-renewable, recyclable/non-recyclable,

---

[96]As Röpke argues in *Civitas Humana,* p. 5: "Feudalism and monopoly are mutually related". See, in particular, the Fourth part on the decentralisation of the industry. The primacy of the small and long term means a rejection of the Stock Market as a cornerstone of the economy. On its part, the bank is seen as a financer of business and, at the same time, businesses are based on the participation of workers. The reform of the financial system proposed by Paul Volcker ran along these same Ordoliberal lines by aiming to eradicate the banks who were too big to fall and dividing them instead. By contrast, the Basel III agreements may make matters difficult for small banks and savings banks by increasing their reserves.

[97]Dembinski, P. H. *Finance Servant*, cit. Ch. 6. 6.

[98]This was the central lesson of H. Bergson's work, that the world of mathematics ignores the true dimension of time and therefore of the human. On this see *Le pensée et le mouvant* (Paris: PUF, 1969), passim.

[99]Naredo, J. M, *Raices económicas del actual deterioro ecologico y social*, siglo XXI, Madrid, 2006.

necessary/superfluous. In Germany, the country with the greatest level of ecological sensitivity in Europe, three basic principles of this sustainability have been emphasised since the 1990s:

1. ecological efficiency, which is to say the best use possible of resources; doing more with less
2. coherence or biomimicry, which is to say the adaptation to ecosystems and
3. sufficiency; not consuming any more than is necessary.[100]

Sustainability must go hand in hand with responsibility both in a synchronous way, with our peers, as well as diachronically with future generations.

Synchronic responsibility demands that the use of both endosomatic and exosomatic energy be distributed fairly, in accordance with others' needs in contrast to the extremely unjust distribution of today according to which 20% of the population consumes 80% of the world's resources.[101] As stated in *Caritas in veritate*: the South needs to grow and the North must put a stop to its growth and its wastage. This is what is happening in effect, given that it is the emerging countries, especially China, India, Brazil, Russia and Turkey who are growing, although unfortunately this growth does not go hand in hand with a fair distribution of wealth but, on the contrary, it further perpetuates inequalities. According to data from the *New Economics Foundation*, the 200 largest corporations manage 29% of the worldwide economic activity but employ only 0.25% of the global workforce. Because of this, it is also important to distribute working hours more fairly as a way out of the current crisis in line with the European Social Charter.[102]

Sustainability is, above all, related to diachronic responsibility, with the duty not to leave our descendents with a world which is worse than the one we inherited. This is related to the aspect of permanence, which has been considered a cornerstone of wisdom.[103] "From an economic point of view, the central concept of wisdom is permanence. We must study the economics of permanence. Nothing makes economic sense unless its continuance for a long time can be projected without running into absurdities".[104]

This permanence is threatened today primarily by the volatility of the markets which is a product of irrationality and of the unlimited desire for profit. It is no

---

[100]Josef Huber, "Nachhaltige Entwicklung durch Suffizienz, Effizienz and Konsistenz", in Peter Fritz, *Nachhaltigkeit in wissenschaftliche and soziologische Perspektiv*, 1995; Josef Huber and Manfred Linz, in their studies, *Weder Mangel noch Ubermass*, (Nieither scarcity nor excess) *Uber Suffizienz and Suffizienzforschung* Instituto Wuppertal, 2004.

[101]Wuppertal Institute *Fair Future. Free Fall, Resource Conflicts, Security and Global Justice* (London: Zed Books, 2007), ch. 5, pp. 158–65.

[102]This is what has been duly carried out in Germany. On the current relevance of Ordoliberal and anti-capitalist ideas, see N. Blüm, *Gerechtigkeit. Eine Kritik von der Homo oeconomicus*, 2007.

[103]Small is Beautiful, cit. 1,4.

[104]Small is Beautiful, cit.

accident that the notion of human security that attempts to eradicate all avoidable suffering would appear precisely to confront particularly difficult situations created by the *downside risks* of finances.[105]

But the establishment of a culture that would restore responsibility would require, above all, a completely new kind of anthropology that would know how to value money correctly as something that is indispensable to overcome certain structural deficiencies, such as poverty, the lack of drinking water, food and pharmaceutical products, but that cannot resolve the problem of power differences between human beings[106] and, above all, can do nothing about the ontological deficiencies of the human being, especially with regards to suffering and death.[107] Because of this, a sense of responsibility can only be recovered, after the playful parenthesis of the last 40 years, by recognising the instrumental, subordinate and limited role of money, a role which forces society to be organised in such a way that it is not brought to its knees by "the markets", but so that, by contrast, finances can be subordinated to production and this, in turn, to the human being.

The absolutisation of the market leads to a negation of the undeniable dignity of the human, by considering that everything has a price, that everything can be bought or sold. Only by overcoming this absolutisation can there be a reestablishment of the ontological dignity of human beings and of their rights, which should not only be considered inviolable by others, but also as inalienable rights of the holders themselves.

On the other hand, it is important to avoid the contradiction which arises between proclamations of the sustainability of the economy, which should take into account our responsibility to future generations, and the legalisation of the murder of those already conceived. How can we worry about the life conditions of our great-grandchildren while the right to kill our children is simultaneously being granted?

---

[105]On this see J. Ballesteros "Sicurezza umana e indivisibilità dei diritti", *Ragion Pratica,* 2008 (30), p. 13–26.

[106]On the irreducibility of the subject of human capacities to simple economic needs, see the work of A. Sen, *Development as Freedom* (Oxford: Oxford University Press, 1999). Sen is also the main architect of the theory of human security. On the problem of the disabled, see article 3 of this book.

[107]This recalls the evangelical passage Lc.16,15 on the uselessness of the accumulation of riches in the face of death. On this, see J. Ballesteros, "Mas alla de la eugenesia. El posthumanismo como rechazo de la idea del *homo patiens*" in *Cuadernos de Bioética,* 2011/2, pp. 13ss. E. Becker in *The Denial of Death* (New York: Simon & Schuster, 1973), already stated that the lack of a sense of transcendence is the greatest problem facing economic equality.

Chapter 2
Basic Structure and Tax Havens

Francesco Biondo

Abstract Chapter 2 argues that the Rawlsian Law of People is much more useful to the debate on international distributive justice than is commonly assumed by many leading political philosophers (Pogge, Nussbaum, Sen, Beitz, Singer). I will show:

1. that these criticisms are misleading insofar as they do not take the concept of basic structure seriously and reduce LP to an instance of ethical statism or acritical acceptance of cultural pluralism;
2. that the basic domestic structures, and the principles that govern them, depend on the law of peoples and also its principles. Therefore it is not true that each basic structure is independent of what occurs in the international field;
3. that, even though his argumentation is not clear, Rawls admits the use of distributive principles in the international domain, but only to some extent. Therefore, it is not true that his doctrine has no normative bite. On the other hand, the discussion has focused only on the application of the difference principle as the sole principle that should be applied even in the international domain, on the basis that it nullifies the influence of arbitrary factors such as birth in poor countries. However, as I will show later, this position assumes what Rawls has denied in many passages of A Theory of Justice: that his doctrine is a form of luck egalitarianism;
4. that the concept of basic structure as a set of institutions that can govern the entrance and exit of their members and their wealth is necessary in order to provide a normative evaluation of phenomena such as multinationals' tax avoidance through the use of off-shore branches situated in so-called tax havens. Far from being suited to a "vanished Westphalian World", Rawlsian doctrine can

F. Biondo (✉)
Philosophy of Law, University of Palermo, Palermo, Sicily, Italy
e-mail: Francesco.Biondo@unipa.it

J. Ballesteros et al. (eds.), *Globalization and Human Rights*, Ius Gentium: Comparative Perspectives on Law and Justice 13, DOI 10.1007/978-94-007-4020-4_2,

offer us moral guidance when we face these forms of free riding that threaten the integrity and stability both of national tax systems and, as we are experiencing right now, of the international economy.

Thus, I hope to show that, if my points are correct, there is room for a theory of international distributive justice that follows the principles formulated in LP.

The publication of the book *The Law of Peoples*[1] (hereinafter LP) has produced a huge debate in the fields of Political and Legal Philosophy. To some extent the long-standing discussion among those who reject and those who accept and defend Rawls' point has confirmed what was claimed by Robert Nozick with the publication of *A Theory of Justice*[2] (hereinafter TJ): "Political philosophers now must work within Rawls' theory or explain why not".[3] In this paper I try to revise part of this great debate starting with one assumption; that Rawls' doctrine is much more useful than has been shown by many of his critics. It is useful in the banal sense that it offers normative principles in order to evaluate and redress unjustified inequalities in power and wealth, not only among countries (in the international domain) but also among citizens inside the same political community.

But in which sense can the feeling of the "uselessness" of Rawlsian doctrine arise? I think that this idea may emerge when we take a quick look at some of the most cited criticisms of LP.

It has been argued that Rawls' theory is a doctrine that is suited to a "vanished Westphalian world", but not to the real situation of international relations in which national states have lost their self-sufficiency.[4] Moreover it has been claimed that LP is morally flawed since it is an instance of a contract theory based only on the principle of mutual advantage.[5] Even T. Pogge, an author that shares common Kantian ground with Rawls, points out that this theory rests on a double standard of sorts, in applying only to the domestic sphere those principles of justice that should be applied even in the international domain, such as the difference principle.[6] Furthermore, Rawls uses a form of explanatory nationalism in justifying the economic success of different peoples. Thus, he overlooks the influences that inequalities in power and wealth among countries often produce within the domestic domain of many poor countries.[7] Moreover, it has been argued that Rawls vindicates

[1] J. Rawls, *The Law of Peoples with "The Idea of Public Reason Revisited"* (Harvard: Harvard University Press, 1999).

[2] J. Rawls, *A Theory of Justice* (1971), rev. ed. (Oxford: Oxford University Press, 1999), hereinafter TJ.

[3] R. Nozick, *Anarchy, State and Utopia* (Oxford: Basil Blackwell, 1974), p. 183.

[4] Allen Buchanan, "Rawls's Law of Peoples: Rules for a Vanished Westphalian World," *Ethics* 110, no. 4 (2000): 702–703, and 716–720.

[5] Martha Nussbaum, *Frontiers of Justice* (Cambridge, MA: Belknap Press, 2007), pp. 248–250.

[6] T. Pogge, *World Poverty and Human Rights. Cosmopolitan Responsibilities and Reforms* (Oxford: Polity Press, 2002), pp. 110–115.

[7] Ivi, pp. 139–145, see also Nussbaum, cit. pp. 239–240.

an overly short, or *ad hoc* list of basic rights, so that its doctrine ends up tolerating massive violations of human rights in the domestic domain and affirming the superiority of the people over the individual.[8] Finally, he was criticised for not addressing in full the existence of a global basic structure that pervasively affects individuals of diverse countries,[9] and for not recognising the moral duties of rich people towards poor ones as duties of justice, and not as a matter of charity.[10]

In order to argue in favour of the usefulness of Rawlsian doctrine, I will show:

1. that these criticisms are misleading, insofar as they do not take the concept of basic structure seriously, and they reduce LP to an instance of ethical statism or of acritical acceptance of cultural pluralism;
2. that the basic domestic structures, and the principles that govern them, depend on the law of peoples and its principles as well. Therefore it is not true that each basic structure is independent of what occurs in the international field;
3. that, even though his argumentation is not clear, Rawls admits the use of distributive principles in the international domain, but only to some extent. Therefore, it is not true that his doctrine has no normative bite. On the other hand, the discussion has focused only on the application of the difference principle, as the only principle that should be applied even in the international domain on the basis that it nullifies the influence of arbitrary factors such as birth in poor countries. However, as I will show later, this position assumes what Rawls has denied in many passages of TJ: that his doctrine is a form of luck egalitarianism;
4. that the concept of basic structure as a set of institutions that can govern the entrance and exit of their members and their wealth, is necessary in order to provide a normative evaluation of phenomena such as multinationals' tax avoidance through the use of off-shore branches situated in the so-called tax havens. Far from being suited to a "vanished Westphalian World", Rawlsian doctrine can offer us moral guidance when we face these forms of free riding that threaten the integrity and stability both of national tax systems and, as we are experiencing right now, of the international economy.

Thus, I hope to show that, if my points are correct, there is room for a theory of international distributive justice that follows the principles formulated in LP.

---

[8]See Buchanan, cit., pp. 697–698, 716–720; C. Beitz, "Rawls's Law of Peoples," *Ethics* 110, no. 4 (2000), pp. 686–687; Cecile Fabre, *Justice in a Changing World* (Oxford: Polity Press, 2007), pp. 104–105; S. Caney, *Justice Beyond Borders. A Global Political Theory* (Oxford: Oxford University Press, 2007), pp. 81–84.

[9]It has been argued that Rawls claims the self sufficiency of states, something that is clearly contradicted by empirical and historical evidences. Buchanan, *Justice, Legitimacy and Self Determination*, cit., pp. 209–210, and 212–216. See also, P. Singer, *One world, The Ethics of Globalization* (New Haven: Yale University Press, 2002), pp. 9–13, Nussbaum, cit., pp. 234–237.

[10]A. Buchanan, *Justice, Legitimacy and Self Determination. Moral Foundations for International Law*, cit., p. 215.

2.1 In What Sense Does Justice Depend on Basic Structure?

The concept of basic structure of society is present in Rawls' oeuvre as a whole, since it forms the "topic" of social justice.[11] It is worth citing the author at length:

> The primary subject of justice is the basic structure of society, or more exactly, the way in which the major social institutions distribute fundamental rights and duties and determine the division of advantages from social cooperation. By major institutions I understand the political constitution and the principal economic and social arrangements. (...) The basic structure is the primary subject of justice because its effects are so profound and present from the start. (TJ 6–7).

Examples of these major institutions are the norms that assign basic rights, and the property of means of production, the competitive markets, and the monogamous family. Therefore, if we want to know whether or not a society is just we must subject its basic structure to scrutiny and apply to its institutions the principles of justice that would be agreed upon in the original position. Moreover, the basic structure, through the application of the two principles of justice, frames the transactions and agreements that occur in society in a way that over time preserves background justice "for all persons equally, whatever their generation or social position" (JF, p. 54). Finally, the basic structure has an educational function, insofar as it educates citizens to "publicly exhibit and encourage the ideal of political justice", necessary in order to foster their acquaintance and participation in the public culture. Hence the basic structure is an arrangement that simultaneously "satisfies given desires and aspirations" and "arouses further desires and aspirations in the future" (JF, p. 56).

We can then summarize these points by attributing four features to the basic structure: (1) it is pervasive, (2) it has a coercive nature insofar as it deploys justified (through the application of the two principles of justice) authority towards all citizens, (3) it maintains background justice over time, (4) it forges citizens' preferences in order to ensure the stability of social cooperation and realize the ideal of political justice.

On the other hand, the concept of basic structure does not apply either to private associations and other "less comprehensive social groups", or to the relationships between different societies at the international level. As Rawls clearly states, the principles that apply to the basic structure "do not hold for all cases" (TJ, p. 7).

The rational for this exclusion is twofold. Associations and communities cannot be considered part of the basic structure since they are based on voluntary ties from which one can free oneself whenever one wants, albeit with an high emotional or monetary price. In other words one always has the right to exit, while in the political associations one does not have such a right insofar as we must assume it as a "closed society: that is, we are to regard it as self-contained and as having no relations with

[11] See also J. Rawls, *Political Liberalism* (New York: Columbia University Press, 1993), pp. 11–12 (hereinafter, PL); id. *Justice as Fairness. A Restatement*, ed. Erin Kelly, 55 (Cambridge, MA: Belknap Press of Harvard University Press, 2001) (hereinafter JF).

other societies. Its members enter it only by birth and leave it only by death" (PL, p. 12). Of course this is an abstraction, but it is an important heuristic tool, at least in this case, insofar as it shows a difference between political relations and other ties (such as the participation in a club or in a church); namely that we cannot evade our obligations towards our fellow citizens, at least not in the same way we can do when we take part in a private association. This is clear in the following passage of JF: "We are born into society, and while we may be born into communities also, into religious and their distinctive cultures, only society with its political form of government and its law exercise coercive power. While we can leave communities voluntarily (the constitutional liberties guarantee this: apostasy is not a crime), there is a sense in which we cannot leave our political society voluntarily" (JF, p. 20, see also pp. 93–94). The basic structure corresponds with what could be called "political community" not only because of its pervasiveness, but also because of the peculiar way in which rewards are distributed: "a community can reward or single out its members in proportion to their contribution to its shared values and ends; but a democratic society has no shared values and ends (falling under the good) by which its citizens can be distinguished. All who can be fully cooperating members of political society count as equals and can be treated differently only as the public political conception of justice allows" (ivi, pp. 20–21, see also PL, pp. 278–280). The nature of these special bonds becomes clear if we take a look at the consequence of emigration. We can emigrate, and therefore a possibility of exit from a political community exists, but it is not possible to exit from a political relationship in the same way as we leave an association or a community; in this sense we are not "free" to accept political authority (JF, p. 94). The point is not only that our political identity and the values we share depend on the culture and history of the country in which we have grown up, i.e. the basic structure in which we have been politically educated (JF, pp. 181–182). What is relevant here is the circumstance that we leave a political community in order to enter another one, with new authorities and new duties to maintain local just institutions. Moreover, we may have a right to exit, but, at least in an ideal situation in which each nation is free and equal to others, and individuals enjoy basic civil and political freedoms, we do not have the right to be accepted in other political communities.[12] This is so, because each people is

[12]In LP the topic of immigration is considered a consequence of the persistence of non-ideal conditions in the international domain, such as massive violations of human rights, extreme poverty, lack of access to international markets in fair terms. As soon as the ideal conditions are met, the question of immigration ceases to be relevant. LP p. 18. M. Nussbaum criticizes Rawls for not taking into account among the causes of emigration the economic inequality among peoples (cit. p. 239) and the discrimination of minorities (ivi, pp. 253–254). But she overlooks two passages in Rawls' line of reasoning, which will be introduced with more details later; economic inequality in the international arena is morally acceptable if it is not too extreme and if it is the result of policies pursued by a well ordered society. Moreover, according to Rawls, the rights of minorities are respected in a decent society. These points are, in my opinion, overlooked by Eric Cavallero, "An Immigration-Pressure Model of Global Distributive Justice," *Politics, Philosophy & Economics* 5, no. 1 (2006): 115; Fabre, cit., pp. 116–124.

responsible for the welfare of its citizens, and it is up to them to make efficient use of the resources of its land in order to ensure the population a social minimum and satisfy their basic needs.[13]

So far we have shown why the basic structure as a subject of justice is different from less comprehensive associations. But what about the relationship among diverse basic structures? Are these relationships governed through a "global basic structure"?

Rawls affirms that "once we have a sound theory for this case [namely domestic justice], the remaining problems of justice will prove more tractable in the light of it" (TJ, p. 7). It seems then that the questions related to international justice can be faced only after having solved the questions of domestic justice. But does this mean that there is nothing analogous to the basic structure in international relations? In LP, Rawls seems to offer two distinct positions. From one side he affirms that it is not possible to envisage something analogous to the state at world level (LP, pp. 36, 48). Hence, it is not possible to speak of the existence of a global basic structure, at least one analogous to what we find in the domestic domain.[14] On the other hand, in other passage, he recognizes the existence of a "basic structure of the Society of Peoples" that is the topic of a doctrine of just relationships between peoples (LP, p. 61).[15] Of course, this basic structure is not something analogous to the world state, but it has the same pervasive effect on domestic policies. This point is, in my opinion, clear, when in JF Rawls affirms: "observe though, that beginning with the justice of the basic structure does not imply that we cannot revise our account for a democratic society (domestic justice) in view of what justice between peoples turns out to require. The two parts of a more complete political conception – the justice of domestic society as well as of the relations between societies – can be adjusted to each other in the course of working them out" (JF, pp. 13–14). There is then no "mechanical" or "syllogistic" relations from the principles applied in the domestic level and those to be followed in the international relations, but rather we can speak of a sort of reflective equilibrium even among the principles of justice governing the

---

[13]LP, sec. 1.3 and 4.3. See also D. Miller, *National Responsibility and Global Justice* (Oxford: Oxford University Press, 2007), pp. 244–247. The question of immigration opens up difficult issues, such as which position to take when a dictatorship, for reasons of economic and social crisis, decides to let, or incite, part of their population to leave the country, in order to force the neighbouring rich countries to help the government in need. This occurrence happens in Albania in the summer of 1991.

[14]The analogy of the basic structure with the state is affirmed by S. Freeman, *Justice and the Social Contract* (Oxford: Oxford University Press, 2007), pp. 268–270. He also affirms that there is nothing analogous to a basic global institution: "The rules and institutions that make global economic cooperation possible are national, and they apply internationally only due to agreements among peoples", ivi, p. 307, an assumption that I will consider below.

[15]This point is often overlooked by those who criticize Rawls for his denial of the existence of a basic structure. See as an example Buchanan, "Rawls's Law of Peoples: Rules for a Vanished Westphalian World", cit. p. 705, and id. *Justice, Legitimacy and self determination*, p. 214. However, as we will see later, the existence of a basic structure does not suffice in providing an argument in favor of the application of the difference principle in a global scale.

diverse pervasive relations.[16] Hence, the principles of domestic and international justice influence one another. Moreover, Rawls recognizes that people can build up "cooperative organizations and agree to standards of fairness for trade as well as certain provisions for mutual assistance" (LP, p. 42, and sec. 16.1). Thus, there is a sort of social cooperation even in an international domain that is ruled by the basic structure of the "society of peoples".

For what I have shown thus far, it seems that Rawls is caught in a contradiction insofar as he simultaneously recognizes and denies the existence of a basic structure at the international level.

But are we talking about the same type of basic structure? I think not. This is clear, in my opinion, when we take into account how, since the first edition of TJ, Rawls was aware of the necessity of expanding his theory of justice, from a domestic to an international level (TJ revised edition 331–333).[17] LP offers this extension within the frame of a liberal conception of justice, in order to be able to "work out the ideals and principles of the foreign policy of a reasonably just liberal people" (LP, p. 9). Then, it is not absurd to argue that we should consider the differences between the two types of basic structure that are the object of the two doctrines.[18] Of course, what is lacking in a basic structure at an international level are, at least in this historical moment, the same principles of democratic legitimacy that are applied at the domestic level, and the possibility that coercion might be used in order to implement liberal principles on a global scale. These limitations are, however, coherent with the idea of political legitimacy, insofar as they are due to the tolerance that liberal communities must show towards other communities, if they want to consider them as free and equal members of the society of peoples (LP, pp. 23–25). That is why intervention is not allowed in order to implement all liberal principles of justice over the world.

The question of the justification of decent societies as members of the society of peoples remains open. It is worth recalling that, according to Rawls, these societies are not aggressive, their members are capable of bearing duties, recognizing some human rights, but not all political rights, respecting a common and freely accepted theory of good and justice, and they have a decent consultation hierarchy (LP, chapters 8–9). Liberal societies must show respect to decent peoples, otherwise they would risk imposing their political liberal values on other people (LP, ch. 7) and infringing their "amour propre", the sense of the self respect that each people

---

[16]I believe that this "mechanical" interpretation of the relationship between the principles of justice in the three domains is claimed by Freeman when he affirms that "domestic justice, the justice of the basic structure, has a kind of priority over other forms of justice. (. . .) We can, however, specify the rights, duties and obligations that citizens of a democratic society owe to one another without first working out the terms of the Law of Peoples". Freeman, cit., p. 270.

[17]On this point see Freeman, cit. p. 270.

[18]Thus, we could argue that the characteristics of a basic structure depend, among other things, on the principles that rule this structure. Being diverse the principles applied to a structure from those who are applied to another one, we can say that the two structures are different.

has towards their history and culture (LP, par. 3.3).[19] Thus, the rationale for taking peoples into account, rather than individuals, is not due to the influence of ethical statism, or to an attempt to disguise the Hegelian political philosophy by covering it with a liberal dress.[20] Rawls does not affirm that the individual has a moral role only insofar as he takes part in a political community. In this case it would be absurd to include among the principles of the law of peoples the duty of each people "to honour human rights", "to observe specified restrictions in the conduct of war", or to "assist other peoples living under unfavourable conditions that prevent their having a just or decent political and social regime" (LP, sec. 4.1).[21] He offers a more complicated argument, in which each people is formed by individuals that can choose to participate in the society of people as liberals or as a decent political community, according to their shared history and culture. And this attachment to a culture and to a particular common life is "a good for individuals and associations" that must be preserved if we want to honour the idea of "people's self determination" (LP, p. 111). However, this respect for collective choices does not indicate a lack of concern about the massive violations of human rights of minorities, as has been argued by many scholars, as we have seen at the outset. It is due to the recognition of decent societies "as free and equal peoples and as politically autonomous, capable of self-determination and capable eventually of coming to an acceptance of liberal rights themselves".[22]

The principle of non-intervention in decent domestic policy is in my opinion overlooked by those who criticize Rawls for having offered an overly short or ad hoc list of human rights.[23] These scholars' arguments start with the assumption that there is a correlation, both conceptual and sociological, among rights, so that

[19]The question of paternalistic intervention in international politics has risen dramatically after the US invasion of Iraq. See, A. I. Applebaum, "Forcing a People to be Free," *Philosophy and Public Affairs* 35, no. 1 (2007), 359–400. It is clear, then, how Rawls uses a "normative", rather than sociological or ethnic, notion of "people", in the sense that it is a group of individuals that freely recognise each other as members of a polity.

[20]See Freeman, cit. p. 314. However, it undeniable the deep influence that Hegelian legal philosophy has exercised on the concept of decent consultation hierarchy. See LP, sec. 9.2.

[21]Moreover Rawls does not state that decent and liberal peoples have the same moral values, since the latter are "guilty of injustice". LP, pp. 62, 83. See also Freeman, cit., pp. 299–304.

[22]Freeman, cit. p. 278. For this reason I find the following Nussbaum position too hasty: "Any extension of human rights thinking, by which nations decide to change their structures in response to international debate, is precluded by Rawls' assumptions." Nussbaum, cit. p. 243. She cites as examples the provisions of the Convention on the Elimination of All Forms of Discrimination against Women, acronyms CEDAW. I think Nussbaum misunderstands Rawls. In fact, the Law of Peoples does not reject, as far as I know, the possibility that through a dialogue in the UN, a decent society might accept new legislation that reduces discrimination. The question is whether or not an external intervention for imposing these provisions is justified. This intervention is clearly rejected by Rawls, and by Nussbaum as well (Nussbaum, cit. pp. 255–256).

[23]Kok Chor Tan, "Liberal Toleration in Rawls's Law of People," *Ethics* 108 (1998), 276–295, Beitz, cit. pp. 686–687, Buchanan, *Justice, Legitimacy and Self Determination. Moral Foundations for International Law*, cit., pp. 40–44, 159–176, Caney, cit. pp. 81–82.

the infringements of some rights can produce the reduction of the opportunities of exercising other rights. This interconnection is implicitly admitted by Rawls himself when he affirms the historical relations between democratic institutions and economic development (LP, sec. 15.3). It is a relevant empirical question whether or not a decent society can be stable over time, in the sense that it is not clear which incentives the elites have in hearing the protests of those who do not enjoy full political participation. In this case, decent societies tend to transform into outlaw states, or into instances of benevolent absolutisms, thus increasing the risk of international conflicts. This risk is taken into account by Rawls himself and he ends up stating that if a decent society does not respect human rights, then even these societies may be considered outlaws (LP, sec. 9.1, 10.1).

Nonetheless, the role played by empiric considerations remains open to discussion, especially for a doctrine that claims to be applied to what international relations are, and to the real practises being followed (LP, p. 83).[24] This point is relevant when we scrutinize the Rawlsian claims regarding the set of human rights that must be guaranteed by decent societies. According to Caney, Rawls employs the empiric consideration selectively, while a deeper analysis of the literature would lead him to recognize the interconnection among rights.[25]

But, are we sure that we are talking about the same concept? Here one point must be made. The concept of human rights that Rawls uses does not refer to what we commonly understand in a domestic legal system as "fundamental rights". In fact, the patterns of implementation of the same rights are diverse. While in a domestic legal system the violation of human rights brings about a complex judicial procedure, in the international arena, we do not have the same institutions. Only recently have many countries begun to work out a decentralised system of enforcement; in the sense that they recognize in their legal systems the universal jurisdictions on issues concerning massive violations of human rights. Thus, when Rawls speaks about human rights he is not referring mainly to a set of moral principles, only those which are backed by legal sanctions, that form one of the prerequisites for entering a cooperative society of free peoples, and, in the case of the gravest massive violations, for the justification of military intervention. What seems then to be an unjustified restriction of the set of human rights finds its rationale in

---

[24]The "practice dependent" nature of Rawlsian doctrine has caused many allegations of unjustified acceptance of the status quo, as we have seen at the outset. However, two points can be made. First, as A. James as showed, Rawls' doctrine can be considered a form of constructive interpretation that strives to offer the best normative interpretation of a practice. A. James, "Constructing Justice for Existing Practice: Rawls and the Status Quo," *Philosophy and Public Affairs* 33 (2005), 300–303. Thus, there is space for moral criticism of current practices. Secondly, it has been argued that practical dependent theories analyse questions of justice as problems of reciprocity, and not of some moral claims valid independently of the social relationships, as luck egalitarians affirm. See A. Sangiovanni, "Justice and the Priority of Politics on Morality," *Journal of Political Philosophy* 1 (2007), 1–4. And Rawls rejects a luck egalitarian justification of principles of justice, as we will see later.

[25]Caney, cit., pp. 83–84.

the function that Rawls acknowledges in his Law of Peoples; it is not to realize the same background justice all over the world or the same form of social cooperation (which would imply the duty of each liberal people to enforce the same type of basic structure all over the world), but it aims to work out the principles that govern the relations among free and equal peoples.

Of course, it may be the case that the Rawlsian set of human rights, with their uncertainties about the full recognition of political rights, ends up justifying unstable societies, in the sense that within these societies there is no reciprocity between the citizens and the ruling elites and therefore the people's sense of justice is constantly undermined (the central role of this sense of justice is explained in LP, p. 15).[26] But this does not depend simply on the circumstance in which the whole set of rights enumerated in the Universal Declaration of Human Rights is recognised in a legal system. What is at stake is the maintenance of some background conditions, and among them the general acceptance of the duty to enhance and keep just institutions,[27] something that at this moment is lacking even in many western societies, as the tremendous rise of inequality and the public, and non transparent, bailouts of many financial institutions prove.[28] Rawls is clearly aware of these threats, especially in his last works where he states that in order to maintain a political community in a stable manner it is necessary that citizens acquire a sense of justice that motivates them to cooperate, without exploiting others citizens (PL, 142–143; JF, 126–126). These threats are especially alarming when elections are decided on the basis of fund-raising efficiency (PL, pp. 359–363), and public opinion is increasingly uninformed and prone to manipulation.[29] Therefore, the criticism about the partial acceptance of a list of human rights loses much of its appeal, especially when we face the empiric state of our liberal, apathetic, communities. The allegation of a partial look to empirical reasons in favour of the interconnection among rights is liable to lead to a bitter surprise; that our liberal societies are only "formally" democratic even though we recognise that political and civil rights are interconnected.[30] As we will see later, the matter is that unjust

[26]It is worth recalling that the normative concept of people, as a "competent collective agent", which claims "the respect owed to any competent agent", lies, and fall, with the recognition and guarantee of basic human rights and political liberties of all its members. Applebaum, cit., p. 377; see also ivi, p. 388.

[27]On the influence of the background conditions on the justice of the basic structure, see M. Ronzoni, "What Makes a Basic Structure Just?", *Res Publica* 14 (2008), 208–215, and id. "The Global Order: A Case of Background Injustice? A Practice-Dependent Account," *Philosophy and Public Affairs* 37, no. 3 (2009), 247–249.

[28]See on this issue J. Stiglitz, *Freefall. America, Free Markets and the Sinking of the World Economy* (New York: W. W. Norton & Co., 2010), and especially, pp. 51–52, 80–81.

[29]J. Rawls, "The idea of Public Reason revisited" (1997), in id. *Collected Papers*, ed. by S. Freeman, Harvard, Harvard University Press, 2001, pp. 580–581.

[30]In this case, I think that the normative nature of "people", as a community of citizens that enjoys political rights, is formally applicable to our society, but if we take a more in-depth look, it is not that clear how "free" our citizenry is.

background conditions in the international arena not only harm the poor peoples, but also favour the rise of inequalities within rich countries and the financial crisis that we are facing right now. But, *pace* many of Rawls' critics, as we will see below, this is an occurrence that justice as fairness can take into account.

2.2 Fairness in International Trade and Global Distribution: Beyond Some Common Misunderstandings

Up to now I have presented some arguments that can be offered to those who accuse Rawls of reducing, without sufficient justification, the normative bite of his Justice as Fairness by assuming that a theory of justice can be applied only to domestic basic structures taken as separate, and independent political units.

These points are, in my opinion, relevant when we look at the huge debate about the existence of principles of justice in the international domain. Is not justice as fairness, as meant by the whole Rawlsian doctrine, fit to offer such principles?

I think it is, insofar as his Law of People epitomises, although in a really embryonic way, the relationship between the fields of political justifications. Moreover, Rawls is aware of the necessity of considering the relationships between societies as "cooperative" and not simply based on the pursuit of self interest, and affirms among the eight principles the duty of decent and liberal societies to help peoples in need.

However, it seems that this duty is not a principle of justice; it is rather a matter of charity that is not morally binding on states. The rejection of the application of any principle of distributive justice seems clear since Rawls denies the application of the principle of difference in the international domain. In trying to impose a maxim on the global scale we would face the following dilemma: (1) either we would enforce the same economic policies, which are followed by rich countries, upon those societies that freely chooses not to industrialize, even though they live up the minimum standard in order to escape starvation, (2) or we would accept that rich countries transfer part of their resources to those free, but poor, societies. Furthermore, the difference principle has no cut-off point, in the sense that it is applicable to whatever standard of living is reached by those who are worst off. By contrast, the law of people justifies transfers only insofar as they are necessary in order to help poor countries to set up those legal and economic institutions that make them free and equal members of the society of peoples (LP, sec. 16.2, 16.3).

Does the rejection of the difference principle imply that justice as fairness is unfit to offer principles of distributive justice in the international domain? Not at all. The matter is that many assumptions are not clearly stated.

First of all, it is assumed that justice as fairness is a form of luck egalitarianism, so that the function of the difference principle is to nullify the consequence of the natural lottery.[31] Given that birth in a poor society is a matter of luck, there is a moral case for applying the difference principle in international relations. For the sake of the argument I will not address the question of how to implement this principle; I simply wonder if it is required by Rawlsian doctrine, or not. It seems that yes, at least insofar as the fear of the rich to be exploited by the poor seems to come up even at the domestic level.[32] Then, "if one accepts that justice requires correcting for any significant distributive effects of morally arbitrary factors, as Rawls seems to accept in his design of the original position, then stopping short of a Global Difference Principle cannot be justified".[33] But this argument rests on an error; that the difference principle only has the function to nullify the consequences of brute luck. If this were the case, then it would be a "principle of redress", but Rawls explicitly denies the identity of the two principles: "The difference principle is not of course the principle of redress. It does not require society to try to even out handicaps as if all were expected to compete on a fair basis in the same race. (. . .) Although the difference principle is not the same as that of redress, it does achieve some of the intent of the latter principle." (TJ, pp. 86–87). The difference principle, then, arranges the basic structure so that "these contingencies work for the good of the least fortunate" (Ivi).[34]

Secondly, it is true that Rawls emphasizes, perhaps overly so, the role of public culture and legal arrangements in order to make a case for the relationship between decent peoples and economic development. But this does not mean that he is unaware of the role of unjust business relationships or the existence of exploitation or other forms of blatant inequalities. For this reason, he asks for a form of social cooperation among nations that would preserve for all peoples the equal access to markets on fair terms,[35] and states the duty of each people to help other burdened societies. Therefore, we cannot affirm that, according to Rawls' doctrine, in the international domain there is room only for humanitarian concerns, while principles of justice do not play any role. The question is whether in order to redress these inequalities a world state is necessary, with the same four characteristics of a

---

[31]It is true that Rawls affirms in TJ p. 378 that the principles of the law of nations aim "to nullify their historical fate". But this does not give a sufficient reason to affirm that international distributive justice must follow luck-egalitarianism.

[32]Singer, cit. p. 178.

[33]A. Altman and C. Heath Wellman, *A Liberal Theory of International Justice* (Oxford: Oxford University Press, 2009), p. 130.

[34]For this reason it has been argued that luck-egalitarians misinterpret Rawls' doctrine insofar as they overlook the political or relational nature of justice as fairness. See, S. Scheffler, "What is Egalitarianism?", *Philosophy and Public Affairs* 31, no. 1 (2003), 8, 24–28, Freeman, cit., pp. 118–119, and A. Sangiovanni, "Global Justice, Reciprocity, and the State," *Philosophy & Public Affairs* 35 (2007), 29.

[35]I think that Rawls' doctrine would not accept the practice of many rich countries of subsidizing their products, which then impedes poor countries to compete with them. See on the issue: M. Kurjanska and M. Risse, "Fairness in Trade II: Export Subsidies and the Fair Trade Movement," *Politics Philosophy Economics* 7, no. 1 (2008), 29–56.

domestic basic structure that we have shown before, or not. If it is necessary, according to justice as fairness, then those who argue for the unfitness of Rawlsian doctrine would be right, since he clearly affirms both that a global state would be imposed upon the interest of each people to self-determination, and that we have certain duties to help people in need, but these duties are not the same as the difference principle, even though they are duties of "justice" and not simply a matter of "charity".

But what can we say about the necessary enforcement of the principles of justice at the supranational level? It is true that Rawls himself does not say much about the question of whether it is possible to conceive of global institutions that, although lacking a centralized coercive agency, can implement distributive principles. However two points can be made.

First of all we must distinguish between distributive agencies and distributive effects. A certain institutions, for instance the WTO, can have only regulative functions, i.e. it guarantees that each member of the organization has equal access to the markets of the other states, and in case of interpretive controversy of the Treaty they will be treated equally.[36] But, this equal access has enormous consequences in the distribution of wealth all over the world.[37] Moreover, whatever the aim of distributive justice might be (in the sense that we can discuss for ages what we think is or should be the currency of justice), it cannot be denied that the form of equality we strive for depends on a number of background conditions (such as enormous differences among nations in power, resources and wealth) that can reduce the capacity of each domestic basic structure to satisfy the principles of justice. In this case, even though an institution does not have the power to assign duties or rights to individuals of other countries, it can produce a change in the background conditions that enhance the implementation of certain principles of justice.[38] Let us suppose that the WTO treaties would be amended in order to protect some rights of the workers, for example by banning the use of minors or recognizing the right to have independent trade unions, then the new regulation of the norms that govern the access to the global market would have a massive impact on the standard of living of the most poor workers. Countries with more respectful regulations would

---

[36]I am not arguing that this is the actual function of the WTO, which is infamous for often being the institution that imposes unfair conditions on poor countries vis-a-vis their access to western markets, see Singer ch. 3 as an example, but I would like to point out how a possible reform of this institution, or of certain of its practices, can bring about huge effects in distributing resources among peoples. A more up-to-date and sympathetic evaluation of the role of WTO is offered by D. Moellendorf, "The World Trade Organization and Egalitarian Justice," in *Global Institutions and Responsibilities*, ed. C. Barry and T. W. Pogge, pp. 141–158 (Oxford: Blackwell, 2005).

[37]Of course, the matter is that poor countries do not have the same opportunities to have a fair hearing in the organization, but this does not mean that a reform of this organization cannot produce, at the same time, a better implementation of the (cooperative) principles that these institution should enhance. For this reason, Moellendorf, among many others, proposes that the participation of the least- developed countries in the organization be subsidised. Ivi, p. 150.

[38]See Ronzoni, "The Global order: A case of background injustice", cit., pp. 249–251, and 254–255.

ask permission of the WTO to impose tariffs on those countries that do not respect workers' rights, which would reduce the advantages for those countries and for the companies or multinationals that have decided to set up there in order to save on labour cost.[39]

The same result could occur when a judge applies the norms of an international treaty of customary international law in civil action for torts committed by multinationals. In this case, the foreigners enjoy a right of compensation for the damages produced by the multinational. An example of this way of enforcing human rights, and therefore to redress the result of some extreme inequalities in power and wealth, or the influence of unjust background conditions, is the Aliens Tort Claims Act that has been used in numerous cases. Plaintiffs can use the act "as a means of attempting to compel multinationals corporations to conduct their operations in a socially responsible manner and to secure compensatory damages if they do not".[40] In this case, the decision can have distributive effects, even though the judge has no authority to decide matters of international distributive justice. Thus, the power of the multinationals is reduced since they must respect populations and workers' rights that are recognised in the law of the nation[41] or in the treaties of the United States, even though local authorities are not able, or do not have any will, to enforce those norms. Two famous cases may be cited. The first is Roe v. Bridgeston Corp. in which the U.S. District Court of the Southern District of Indiana ruled "that claims involving child labour are actionable under the Alien Tort Claims Act". Another case is Khulamani v. Barclay National Bank Ltd, "a case filed on behalf of victims of apartheid alleging that they had been harmed by multinational corporations' compliance with apartheid policies" in which the 2nd Circuit Court of Appeals stated that the claims were actionable under the act cited above.[42] Certainly, there

---

[39]In this case we would take into account the issue of the use of workers that do not enjoy any form of labour rights. The result would be analogous to P. Van Parijs' proposal of putting "patriotic tariffs" on those goods that are the result of exploitative labour relationship in countries in which labour rights are not respected. Philippe Van Parijs, *Real Freedom for All. What (if anything) can Justify Capitalism* (Oxford: Clarendon Press, 1995), 231–233. On the issue in general, see Mathias Risse, "Fairness in Trade I: Obligations From Trading and the Pauper-Labor Argument," *Politics Philosophy Economics* 6, no. 3 (2007): 355–377.

[40]Daniel E. Lee and Elizabeth J. Lee, *Human Rights and the Ethics of Globalization* (Cambridge: Cambridge University Press, 2010), p. 211.

[41]"The law of nations, which is also known as international customary law formed by the general assent of civilized nations, is comprised of rules and principles that govern the relations and dealing of nations and of international organizations with each other; as well as some of their relations with persons, whether natural or juridical", Ivi, pp. 212–213.

[42]Ivi, pp. 216–217. Another important case, that has reached an extra-judicial settlement, is John Doe I v. Unocal Corporation. A group of Burmese villagers accused Unocal of having hired army troops in order to secure the construction of a pipeline. These army troops eventually imposed forced labour on people living in that area. John Doe I was killed while attempting to escape, and his wife was raped. Although the exact amount of the compensation is unknown, it is believed that Unocal has paid 30 millions $ in damages. Ivi, pp. 229–230.

are many arguments for dismissing these claims,[43] but the point I want to make here is simply that there is no reason why the current division of labour among diverse governments cannot have, in principle, positive outcomes in terms of distributive justice. So that, it is incorrect to blame Rawls' doctrine that without the enforcement of the difference principle we leave poor peoples unarmed or we are to accept the current unequal relationships. Even regulatory authorities, such as the WTO, or domestic legal arrangements of restorative justice can bring about distributive effects, and redress unjust background conditions.

Secondly, it is true that democratic deficit remains the first problems in justifying supranational institutions, and the European Union is no exception. However, is the requirement of a constant will in maintaining a form of supranational cooperation a necessary and sufficient condition in order to justify the same existence of a political institution or of a legal arrangement in international relations? The answer depends on the conception of the source of political obligation. It is true that Rawls' position, with his emphasis on the interest of each people to self-determination, can be accused of justifying matters of distributive justice *only* by referring to each nation's will to cooperate. Moreover, in a private correspondence with P. Van Parijs, he affirms that processes of economic integration, such as the European Union, are pushed forward mainly by the financial interests of European capitalists and bankers.[44]

But this remark does not make a cooperative attitude among people "unreasonable", in the Rawlsian sense of "acceptable by free and equal member of a society", even though we do not set up institutions that guarantee the same level of political accountability that we enjoy at the domestic level. As Van Parijs argues, some of the consequences of the process of globalization are so pervasive that they cannot be governed only by national governments. Western democracies run the risk of being forced to enter into competition with other governments in attracting capital through a massive reduction of taxes, at the expense of the guarantee of social rights.[45] In this case we could justify regulative supranational authorities that could help in maintaining or enhancing fair patterns of cooperation among people, even though the members of these authorities are nominated by governments and are not chosen by the people.[46] As examples we can cite the authorities in the field of international

[43]Ivi, pp. 219–227. It is not a case that the more progressive trends in interpreting the Aliens Tort Claims Act have been strongly contrasted by the Bush administration. See ivi, pp. 230–231.

[44]John Rawls and Philippe Van Parijs, "Three Letters on *The Law of Peoples* and the European Union," *Revue de philosophie économique* 7 (2003): 7–20, http://www.uclouvain.be/cps/ucl/doc/etes/documents/RawlsVanParijs1.Rev.phil.Econ.pdf downloaded the 1 of February, pp. 7–8.

[45]Ivi, pp. 9–11.

[46]Of course, the government that appoints those people is supposed to be accountable to its citizens. It is true that the same government can blame supranational authorities for the unpopular policies that are taken. However, it is not enough arguing that a certain policy is imposed upon a democratic government by supranational authority in order to consider it illegitimate. Other things must be considered, such as the merits of that policy, whether it is efficacious or fair even though it is rejected by the majority of the citizens, and whether the government does not want to be held

commerce, which aim to ensure the equal opportunity to access to markets, or those who tries to implement the transparency of capital flows, the avoidance of harmful tax competition, or the enforcement of basic norms in labour conditions, This point is especially telling if we are aware of the existence of externalities produced by the working of these economic structure that have pervasive effects in our everyday life. As A. Abizadesh persuasively argues, each cooperative scheme in international relations must face the issue that "the range of persons pervasively impacted by the institutions of the basic structure is not equivalent to the range of persons who participate in a potentially cooperative scheme of social coordination. This is true because of a rather simple fact: every such scheme has externalities, i.e., those who (can) participate are not necessarily the same as those who are affected".[47] Therefore, regulative organizations that are global in scope can be justified insofar as they can effectively solve problems of coordination among diverse governments, as we have seen in the case of tax competition or equal access to market. And this regulation, eventually, can help in maintaining just background conditions and reducing the unequal distribution of bargaining power among rich and poor countries.

2.3 Humanitarian Hypocrisy and Tax Havens

I hope I have succeeded in showing how various replies can offer a certain interpretation of LP. In this last section, my aim is more ambitious. I would like to argue that there is a certain understanding of the notion of closed society that is presupposed when we discuss questions of domestic and international distributive justice. In order to make my point I will introduce a case of "humanitarian hypocrisy".[48] Let us suppose we are the owners of an important multinational

---

responsible of taking unpopular measures, although necessary. Moreover, the people could exercise its right to exit from that institution, and therefore the supranational institution does not exercise any coercion. In this case, I would hasten to say that even though some of the recommendations provided by the institution might be considered illegitimate by the majority of the citizens, they might be morally binding, if the citizenry do not want to leave this institution.

[47] A. Abizadesh, "Cooperation, Pervasive Impact, and Coercion: On the Scope (not Site) of Distributive Justice," *Philosophy and Public Affairs* 35, no. 4 (2007): 343. In this case we would reduce, as Abizadesh urges, the gap between site (which is local) and scope (which is global) of distributive justice; ivi, p. 320–321, 330, and it would not be confirmed the alleged (by him) inappropriateness of Rawlsian doctrine, ivi, p. 358.

[48] The case that has caught my attention concerns the famous rock band U2. Some journalist investigations and a French documentary (La gran evasion) disclosed that the band had set up a financial company hosted in a tax haven. The aim, legitimate as laws stand, was to reduce the taxes that had to be paid in the case that they would have maintained the site of the company in Ireland, which, by the way, has the lowest taxation on corporate earnings. What has produced indignation was the fact that the band was famous, among other things, for the humanitarian commitment of the singer, Bono, who was used, and he still is, to make public exhortation in favour of the duty of

company with its financial branch in a country with very low or no taxation, a very efficacious legal arrangement that ensures total secrecy regarding the ownership and quantity of our assets in those banks, and finally a constant practice of rejecting, or blocking, any form of cooperation with other countries that could shed light on the opaque architecture of financial institutions that are hosted in that territory.[49] Unlike 99% of our fellow citizens, thanks to the work of an international tax law firm, we enjoy the opportunity to transfer the earnings of our business to such a tax haven. Thus, we, or the company we own, is allowed to save a huge amount of money. We thus manage to escape the same fiscal imposition that is enforced upon our fellow citizens.[50] On the other hand, we feel we are obliged to help those people in need in other, far-away countries. We consider their destitution to be something they cannot be blamed for, something that depends on a matter of brute luck. They were born in the wrong society, and we, the well-off, have the moral duty to improve their standard of living because of the fact that we are luckier than others. It is not a matter of charity, we argue, but a "matter of justice". Of course, we are not unconcerned with the situation of those in society who are worse-off, but we are aware of the fact that it is more urgent to help those needy people that suffer absolute deprivation.[51]

---

rich government in helping poor countries. Protesters accused the band of being hypocrites insofar as while they were criticising governments for not doing enough in helping poor people, they were avoiding their fiscal obligations towards their fellow citizens. See, with a short reply of the singer, "Tax protesters rattle U2 with hypocrisy jibes," *Sunday Times*, March 1, 2009.

[49] According to the OECD definition, the "Tax haven criteria" are: "(1) there is no or nominal tax on the relevant income (from geographically mobile financial and other service activities); (2) there is no effective exchange of information with respect to the regime; (3) the jurisdiction's regimes lack transparency *e.g.* the details of the regime or its application are not apparent, or there is inadequate regulatory supervision or financial disclosure; and (4) the jurisdiction facilitates the establishment of foreign owned entities without the need for a local substantive presence or prohibits these entities from having any commercial impact on the local economy". Organization for Economic Cooperation and Development (OECD) *Towards Global Tax Co-Operation: Progress in Identifying and Eliminating Harmful Tax Practices. Report to the 2000 Ministerial Council Meeting and Recommendations by the Committee on Fiscal Affairs* (Paris: OECD, 2000) http://www.oecd.org/dataoecd/25/27/44430257.pdf, p. 10, n. 4, downloaded the 1 of February 2011. A taxonomy is offered by Lorraine Eden and Robert T. Kurdle, "Tax Havens: Renegade States in the International Tax Regime?", in *Law & Policy* 27, no. 1 (2005): 101–102. We must take into account, however, that these characteristics can be realized in different ways and to diverse extents through the particular legal arrangements that are set up to guarantee the interests of those who want to escape high taxation and keep their wealth secret. What is of much interest is the fact that in the list published by these authors there are cited territories that are dependent on, or belong to, those governments that, supposedly, aim to eradicate or reduce the impact of harmful tax competition through the policy recommendation set up by the OECD. Ivi, pp. 111–112.

[50] Gillian Brock, "Taxation and Global justice: Closing the Gap between Theory and Practice," *Journal of Social Philosophy* 39, no. 2 (2008): 165 at footnote 22 cites the case of Microsoft as paradigmatic. Microsoft succeeds in taking huge earnings (12.3 billion $ in 1999) without paying taxes in the US through complex transfers of all earnings into its branch hosted in a tax haven.

[51] I would like to point out that I am not affirming that Bill Gates is taking advantage of illicit practices. I simply wonder whether his philanthropic attitude is coherent with a practice that permits him to avoid paying those taxes that are imposed upon 99% of his fellow citizens. Given

Moreover, we can claim that our fellow citizens are supposed simply to be *only relatively poor*, or not motivated enough to take advantage of the opportunities that our society offers. Be that as it may, we announce publicly, along with other rich entrepreneurs, bankers and manager of funds of investment, the launch of a huge humanitarian foundation that aims to enhance the standard of living of those populations.

What does make this attitude, if it does, an instance of hypocrisy, of manifest contradiction, or worse a case of "moral marketing" ("Look how moral I am! I believe in compassionate capitalism". "Buy these products and you will help the poor countries, the orphans in third world countries.... and so on")?

The matter is that contributions for the functioning of those institutions that guarantee the production of public goods are unfairly allocated.[52] Ninety-nine percentage of our fellow citizens cannot take advantage of these tax havens, because they have a lower, or immobile, tax base. Thus, they face a dilemma; either they have higher taxes, or they get less public services, or with less quality. This dilemma produces a vicious circle whose victims are the poor and those in need of public services. Since the existence of a practice of tax avoidance, though through the use of legal loopholes, such as the establishment of branches in tax havens, undermines the general motivation to comply with tax norms, then the government must increase the pressure towards those who cannot avoid the imposition. As Augustìn Menendez clearly states: "while some tax bases are become increasingly elusive, others have become even easier to control: for them, it is almost impossible to hide from tax authorities. It has been said that this creates new inequalities, as the *many* are taxed more heavily in order to compensate this revenue lost from the *few*,

that those citizens' wealth and earnings are less mobile, or not mobile at all, they cannot avoid taxation. Moreover, as the OECD report points out, they must bear the burden of the possible shift of taxation towards less mobile tax bases, such as labour, property and consumption. OECD, cit., p. 5. Thus, this practice has led to a huge unfairness towards the vast majority of the fellow citizens that cannot enjoy the same opportunity of transferring their earnings in a tax haven. Moreover, it can be argued that with this attitude the state's same possibility of securing its functions, such as maintaining public security and enforcing property rights, is put under threat. And, as we will see below, the richest people take proportionally more advantage of these functions.

[52]Public goods are those goods that can be consumed collectively without reducing their value or their availability. Public security and public health are public goods since everyone can enjoy them without reducing their value. In public economics, it is well spread the idea that taxation is justified insofar as it ensures, through a mechanism of enforcement, the contribution for making the state able to produce these goods. Surely, the set of these public goods is not fixed once and for all; according to the evolution of society, this set may increase, and along with it the burden of taxation. See, A. Menendez, *Justifying Taxes: Some Elements for a General Theory of Democratic Tax Law* (Dordrecht: Kluwer, 2001), pp. 99–104.

increasing the discontent with the tax system (those who pay it do it *only* beacuse they cannot evade)".[53] But it is not only a matter of fairness that should worry us.[54]

We must bear in mind that the function of these goods cannot be reduced only to the necessary guarantee of public security and property rights through the use of legally justified coercion. The availability of these goods is, often, a necessary condition in order to give citizens the opportunity of consuming, without spending a certain part of their wages in, often inefficient, health insurance schemes. Therefore, it is often the case that the availabity of public goods through an efficient public service can foster economic development. And our revenues depend, among other things, on the possibility to find people that are the beneficiaries of these development. But, as it is clear, the availability of these goods, that is instrumental to economic development, is put in danger when the governments find so many difficulties in collecting through taxes the necessary resources. As Sandra Feldman states:

> Markets are known to function better with the input of public goods, not just in the form of transport, infrastructure, and information technology but also in training, health provision and education. Yet competitive markets themselves do not provide such public goods: indeed globalization has led to a fiscal squeeze, which in turn has made it difficult to for government to provide the very goods necessary for a well-functioning market.[55]

Thus, Rawlsian idea of the closed nature of domestic basic structure can be interpreted in a different way than that which has been offered by many scholars. It is not due to the acceptance of the idea of self-sufficiency of national state. It is instead due to the assumption of the responsibility of each people to keep their wealth, and implicitly, to enforce the duties of the citizens to maintain just institutions. Thus, it is

---

[53]Menendez, cit., p. 66.

[54]As it was put clearly by Hobbes, fairness in allocating the burdens of taxation is of paramount importance in maintaining the "commonwealth" in peace. "The second factor, we have shown, which disposes men to sedition is the discontent that arises from poverty; even if their poverty results from their own extravagance or idleness, nevertheless they blame it on those who govern the commonwealth, claiming they are oppressed and exhausted by taxes. Yet it may sometimes happen that their complaint is justified, namely when the burdens of the commonwealth are imposed on citizens unequally. For a burden which would be light if all shared it, becomes heavy, and indeed intolerable, for the rest when many wriggle out of it. And it is not so much the burden itself that men object to as the inequality. The most contentious struggles are about tax exemptions and in those struggles the less successful envy the more successful as if they had been defeated in battle. It is in the interest of the public peace to remove a justified complaint, and consequently it is a duty of sovereigns to ensure that public burdens are equally borne. Besides, since the citizens' contributions to the community are simply the price they pay to purchase peace, it is logical that those who equally enjoy the peace should pay equal shares, by contributing either money or service to the commonwealth. And the natural law (by III, 15) is that in assessing others' rights every man should treat himself as equal to every other man; hence sovereigns are obliged by natural law to impose the burdens of the commonwealth upon the citizens equally". T. Hobbes, *On the Citizen*, ed. and trans. Richard Tuck and Michael Silverthorne (Cambridge: Cambridge University Press, 1998), p. 147.

[55]Sandra Feldman, *Human Rights Transformed. Positive Rights and Positive Duties* (Oxford: Oxford University Press, 2010), p. 42.

worth bearing in mind that societies must be imagined as closed because we cannot escape our duties to cooperate with other members of our political community.[56] We have seen at the outset, how Rawls distinguishes between political bounds and other associations. This distinction should shed light on the duty of each citizen to pay a fair amount of those revenues that are possible only thanks to the functioning of just institutions. Given that the market transactions are possible only in the legal framework provided by a government, there is then a prima facie duty to cooperate in the maintenance of this institution. But, using the loopholes that many legal taxation systems offer, we can enjoy the benefits of a system of cooperation without paying the costs (or paying only an unfair share of these costs) of its maintenance.

I am not arguing that citizens have no right to emigrate or to export any part of their wealth without the government's permission. What I am saying is that legitimate governments, even those who are only decent, in Rawlsian terms, are justified in controlling and taxing the exit of this wealth. This tax must not be meant as a sanction or an incentive not to abandon a country, but as a contribution to the maintainance of those institutions that have offered the legal arrangements necessary to realize those earnings.

The concept of "closed" basic structure then sheds lights on the question of reducing the individual incentive to "free-ride", to enjoy the benefit of social cooperation but to avoid the cost of maintaining the institution necessary to maintain this cooperation. But it is not only a problem of the enforcement of some norms. Rather, it is a question of the type of relationship that citizens have towards social institutions. One of the most famous taxonomies about the different types of relationship between an individual and an organization is the one offered by A. Hirschman.[57] He distinguishes between exit (which is the behaviour that is put into practice normally in the commercial relationship by the client towards the provider of a good in a free and open market) and voice (which is the behaviour of the citizen that protests, calling for an enhancement of some policies or a change in the legal system). The citizen, unlike the client, exercises the voice because he feels himself to be a member of the organization he criticises, and therefore, if he wanted to exit the organization, he would pay a high price.[58] Political relationships, then, are different in kind from commercial ones, insofar as loyalty to the organization

[56]Thus, my interpretation of the "closed nature" of basic structure does not depend on the respect of the identity of the people that is governed by that basic structure. According to Miller, Rawls' points about the "closed nature" of the domestic basic structure are due to his recognition of people's common identity as a necessary condition for the effectiveness of any redistribution. See D. Miller, *On Nationality* (Oxford: Oxford University Press, 1995), pp. 90–93. By contrast, I offer a less "identity-based" (and more "politically-based") interpretation of this characteristic of the basic structure. On the other hand, I would like to point out that these interpretations are not mutually exclusively, insofar as Miller's interpretation is compatible with my emphasis on the duty of each citizen not to avoid domestic taxation. What I do not agree with is his point on the necessity of a shared identity for the success of a redistributive policy.

[57]Albert Hirschman, *Exit, Voice and Loyalty* (Cambridge, MA: Cambridge University Press, 1970).

[58]Ivi, pp. 15–18, 30–32.

is dominant over the pursuit of the personal interest, provided that the citizen is still convinced that it is useful to exercise his voice instead of simply leaving the organization.[59] Furthermore, public goods can be produced only when the recipients want to keep them available, even though this provision implies some costs. Thus, people are willing to exercise their voice, because they know that their welfare depends on others' welfare and attitudes. Hirschman provides a clear example for explaining the citizen's attitude towards public goods. The example is the choice of the school for the children. The parents have two alternatives: either put the children in a private school, or keep them in a public one, with a lower standard. Besides the monetary cost, the first alternative implies that the parents know that they will not cooperate in maintaining the public good that is the public education, and that sooner or later the children will suffer, indirectly, from the deterioration of that public good. By contrast, the second alternative entails a bigger involvement in the maintenance of that public good. Thus, exercising voice is a form of expressing loyalty towards public institutions or towards private organization.[60] However, the parents do not know with certainty that their voice will be heard or that their commitment will make any difference in the maintenance or enhancement of public education.[61]

In contrast, exit is the attitude common in relationships based on self interest, such as those that we experience in the market. And, according to Hirschman, this attitude forms part of the dominant ideology of the US in the 1960s. According to this ideology, which provided the self-understanding of the vast majority of American people in the 1960s, there is always a territory that offers opportunities to those who want to grasp them. The difference between exit and voice corresponds to two opposite attitudes towards social issues. Those who exercise the exit believe that there is always a solution outside the community and its restrictions. Moreover, there is virtually no public good, but only resources that can be sold and bought in a market. Finally, everyone can achieve the success and prosperity only through his personal efforts. Those who exercise the voice, and among them Hirchman himself, recognise that the solution to social issues depends on the possibility of setting up forms of cooperation with the other citizens. Thus, one's welfare relies on the availability of public goods.[62]

Hence, it is possible to offer an alternative interpretation of "closed" basic structure, or "closed society". This notion has the function of shedding light on the cost of entering and leaving a community, and the reason why it is sometimes necessary to tax the exit of the citizens' wealth. The rationale is to ensure fair

[59] Ivi, pp. 36–41.

[60] In the preface to the Italian edition of this book, Hirschman states that one of the sources of the essay was to be found in his personal experience as a Jew fleeing from Nazi Germany. The feeling of loyalty can produce a form of paradox: the worse is the situation the more one person feels obliged not to leave the country in order to give their contribution to the community. But, the worse the things were going, the more difficult it was to leave the country, even if one were willing to. See also ivi, pp. 98–100.

[61] Ivi, pp. 100–105.

[62] Ivi, pp. 106–112.

terms of cooperation among the citizens, provided that the inequalities in wealth and power can bring about inequalities in taking advantage of legal loopholes in the tax system. Thus, the notion of closed society illuminates complicated forms of free-riding that are even more pressing in times of fiscal crisis.

However, one might argue about what the point is of this discussion for the topic of international distributive justice. One might agree that it is morally incoherent to ask the government to help poor countries and, at the same time, avoiding taxation in his country through a complex financial architecture of branches hosted in tax havens. We can even say that the situation is analogous to that depicted by T. Pogge when he accuses western public opinions of taking advantage of moral loopholes.[63] Like affluent societies who enjoy the natural resources that western corporations buy to corrupt dictatorships, few very rich people can take advantage of the opacity of the tax system, and of the (collectively harmful) competition among governments in collecting capitals. At the end of the day, even if my interpretation is useful, it is of no help in the discussion about the duties that rich countries have towards poor ones.

By contrast, I think that the sense of closed society, or "closed basic structure" I have provided, and what has been said concerning the tax havens, could be useful for the cooperative society of decent and liberal peoples that Rawls has in mind.

First of all, because it has been shown how the principles of domestic and international justice are mutually influential, in a way that it is seldom cited in the debate. The existence of unrestricted circulation of capitals can produce great damages, as the OECD report shows, insofar as it weakens the capacity of each government to collect the resources necessary for providing citizens with the public goods to which they are entitled. It therefore seems to me that a theory that takes into account only the damages towards poor countries that the current lack of international regulation of capital flows can provide, while this lack is harmful towards the middle and low classes in the affluent societies as well, is insufficient.

Secondly, the case for regulating and eventually eliminating the loopholes in tax systems offered by the tax havens is evident in issues of aid to poor countries, or countries plagued by corrupt dictatorships. We have seen that one of the characteristics of the tax haven is the total opacity in financial operations present in those territories as well as the strict compliance of bank secrecy. As many NGO's[64] have denounced, bank secrecy allows dictators, members of military juntas or even powerful criminal organizations to hide their wealth in the banks and holdings hosted in these territories. Thus, we see that citizens of both poor and affluent countries suffer the consequences of the opaque circulation of capital. I therefore

---

[63]The example provided by Pogge concerns the owner of a building who wanted to make more profit from his estate, by evicting the old tenants and building up news luxury flats. However, he did not have the courage to face the reaction of the old tenants, so he hired a private company that will do the unpleasant job. Even though others carry out the job, the owner is nonetheless responsible insofar as they act for his profit. Pogge, cit. pp. 75–80.

[64]See, as examples, Global Financial Integrity (http://www.gfip.org/) and Tax Justice Network (www.taxjustice.net).

share Gillian Brock's arguments about the necessity of elaborating principles of international taxation that could reduce the opportunities of the richest people to avoid, or unjustly reduce, their tax bases in their countries of residence. Hence, the duty of each government to implement an international arrangement that enhances transparency and cooperation among governments in the control and taxation of capital flows has two recipients: the citizens of the affluent societies that cannot avoid taxation, or at least have less opportunities to reduce their tax bases,[65] and the citizens of poor countries that are robbed of their natural resources by corrupted elites. Moreover, this regulation can be considered a precondition for implementing Pogge's proposal of the Global Resource Dividend, since it reduces the opportunities of dictators or military juntas to hide wealth obtained by corruption or through selling the natural resources of their own country.[66] Thus, I hope I have offered one sense of the notion of closed domestic basic structure as compatible with the endeavour of elaborating a doctrine of international distributive justice.

---

[65]As examples of lobbies against tax haven see Business and Investors Against Tax Haven Abuse at www.businessagainsttaxhavens.org; American Sustainable Business Council at http://asbcouncil.org; Business for Shared Prosperity at http://businessforsharedprosperity.org; Wealth for the Common Good at http://wealthforcommongood.org.

[66]Brock, cit, pp. 162–164. The proposal of Global Resource Dividend can be found in Pogge, cit. pp. 204–214.

Chapter 3
Human Rights and the Inclusive Society

Jorge Cardona Llorens

Abstract Chapter 3 approaches disability as a reality that may impede the universality of human rights given that the rights of the group in question are "systematically violated". To achieve an inclusive society in which everyone's rights are implement, it is necessary to start with an adequate definition of the problem, referring to these individuals as "people with different capacities", recognizing the place this difference has in society, and the role the State plays in the integration of this difference. To do this, their social visibility needs to be promoted as well as their accessibility to all spheres of social life, guaranteeing their participation in the labour and political fields, overcoming physical and cultural barriers, and putting an end to the existing confusion between mere social protection and the recognition of rights.

3.1 Human Rights and the Diversity of Abilities

We all have different abilities. Some people have greater visual, auditory or sensory abilities than others. Others are faster, stronger, better with their hands, have a more photographic memory or are better with abstract concepts than the rest. However, although it is clear that these differences exist, it is also clear that in the predominant group in any society, the group that designs and constructs the society's social, political and economical structures, the differences in ability between its members, such as cultural or ethnic, are not usually very big. This means that those with varied abilities (such as those from very diverse ethnic groups, cultures, etc.) become

J. Cardona Llorens (✉)
International Law, University of Valencia, Valencia, Spain

Committee on Rights of Children, U.N. Genevra, Geneva, Switzerland
e-mail: Jorge.Cardona@uv.es

J. Ballesteros et al. (eds.), *Globalization and Human Rights*, Ius Gentium: Comparative Perspectives on Law and Justice 13, DOI 10.1007/978-94-007-4020-4_3,

"different" from the dominant social group, creating the possibility of and the need to avoid discrimination, in order that everyone may enjoy equal rights.

Regarding ability, it has been a historical constant in most societies that people with physical, sensory, psychological or mental limitations which prevent them from reaching the standard considered "normal" by the majority are categorised as "disabled", "deficient", "incapable" and are, traditionally, "protected"; separated from society and put simply, marginalised.

As I hope to demonstrate in this article, this situation has begun to change, first in terms of social mentality and later in law. In the last third of the twentieth century and the beginning of the twenty-first, diversity of ability has begun to be seen, not only as a cause for discrimination, but also as one more of the elements that characterise human diversity and therefore enrich society and as such, should be granted greater importance.

Social enrichment and appreciation, must not only refer to diversity of ability, but to all human diversity: race, gender, age, nationality, etc. The first human right should be the right to be the way we are, in other words, different from others. Equality must not refer to people, but to the exercise of one's rights. We are all different but we have the same rights, and society must establish mechanisms to eliminate barriers which, due to real and specific differences between people, may impede the ability of some to exercise all their rights equally.

We are far from the universal recognition of these values, especially in terms of their full establishment within the law, but nevertheless, there is a move towards a general recognition of the enriching nature of diversity and of the need to guarantee the ability of every member of society to exercise their rights equally, in integrated and inclusive societies.

In order to analyse this evolution I am going to concentrate on people's diverse abilities. Although I feel the term "diversity of ability" is more accurate, in this article I will be using the term "disability" as it is the term most widely used today.

In any case, I should start by saying that although the concepts of human rights and disabilities are deeply interconnected, they have not been studied in any great detail until recently.

The relationship between these two concepts has a double direction; On the one hand, violations of human rights lead to disability; On the other hand people with disabilities are a group particularly vulnerable to human rights violations.

According to the United Nations, at least a third of people with disabilities acquired that disability as a result of a violation of human rights. If we bear in mind that there are an estimated 650 million people with disabilities, this means that more than 200 million people have a disability resulting from human rights violations.

Although this is not the central theme of this address, I do feel it necessary to highlight some of the human rights violations which most commonly result in disability:

(a) First, a whole series of practices, punishments, and forms of torture anchored in religious beliefs, traditions or other cultural factors which have produced a large number of disabilities: amputations as punishment, castration and female genital mutilation, blindfolding and even blinding of prisoners, etc.

(b) Secondly, even more noticeable is the number of disabilities caused as a result of the violation of the human right to normal development. Hunger and malnutrition are the basis for the gradual deterioration of mental and physical abilities. The World Health Organization estimates that one in five people with a disability throughout the world (that is to say, around 125 million people) are disabled as a result of inadequate nutrition. In addition, there are factors such as the lack of adequate sanitation, the deterioration and degradation of the environment, extreme poverty or working in inhumane conditions, all of which increase the numbers of disabled people world-wide, as well as the deterioration of conditions for those already living with a disability.

(c) Third, the area in which the relationship between the violation of human rights and disability is most noticeable is, without doubt, in war zones and conflict situations. In these cases, we are not only talking about physical or "visible" disabilities, but must also include all those affected (many irrevocably) by mental problems and psychological damage.

We can, therefore extract an *initial conclusion* about the relationship between disability and human rights; the violation of human rights and of international human rights law is one of the most important causes of disability. The *second conclusion* is therefore that respect for human rights and international human rights law is an effective way to prevent disability.

However, without denying the absolute importance of this phenomenon, as mentioned previously, my main focus in this article is aimed at addressing the human rights of people with disabilities.

In this respect, we must be aware that those whose abilities differ from those considered "normal", either as a result of a physical, psychological, mental or sensory deficiency, are not a group which is "hypothetically vulnerable" to damage, but rather a group whose human rights are "systematically damaged or violated".

As the United Nations Committee on Economic, Social and Cultural Rights States, in general comment number 5: "Both de jure and de facto discrimination against persons with disabilities have a long history and take various forms. They range from invidious discrimination, such as the denial of educational opportunities, to more "subtle" forms of discrimination such as segregation and isolation achieved through the imposition of physical and social barriers.....Through neglect, ignorance, prejudice and false assumptions, as well as through exclusion, distinction or separation, persons with disabilities have very often been prevented from exercising their economic, social or cultural rights.....on an equal basis with persons without disabilities. The effects of disability-based discrimination have been particularly severe in the fields of education, employment, housing, transport, cultural life, and access to public places and services."

To this we must add the traditional violations of the right to freedom (admission to institutions without minimum judicial process), political rights, particularly the right to active and passive suffrage (complete prohibition or impeding voting);

the right to family (prohibiting marriage, adoption, etc.) and privacy; rights of defence (denying the deaf and blind to a legal defence due to a lack of interpreters, considering illiterate people incapable of defending themselves); without mentioning the problems related to the right to life in cases of extermination of newborns (above all in countries practising demographic control), selective withholding of medical treatment, legislation on euthanasia or the controversial abortion for reason of euthanasia.[1]

Why does this situation occur? In my opinion it occurs mainly due to the existence of three types of barriers:

(a) A principal cause of discrimination has come to be euphemistically known as "*cultural barriers*".

When we speak of cultural barriers, we are talking about a group of attitudes based on tradition, superstition, false suppositions, prejudice, and feelings of shame or rejection towards persons with disabilities. Nobody can deny the existence of these attitudes, or their discriminatory nature, given that they exist throughout the world and create one of the main obstacles to the integration and full participation of people with disabilities in society.

People with disabilities have been marginalised in almost all cultures, throughout history. Common reactions, both by society in general as well as by the authorities have been either compassion or repugnance. The difference that disability represents is perceived as a reason for exclusion instead of a reason to celebrate the diversity of the human family.

Across the world, there are examples of the segregation created by these attitudes stemming from people who do not want people with disabilities in the workplace or near their children in the educational arena. Some cultures consider disability to be a punishment or a curse designed to make the family pay for the acts of their ancestors, further deepening social exclusion. Other cultures limit the rights of people with disabilities in very different ways, including marriage, reproduction, etc. with no justification other than tradition.

(b) Along with these cultural barriers we must also look at a second barrier produced by the *confusion between social protection and the recognition of fundamental rights*. It is thought these people should be *protected* instead of giving them *rights*. This means that these "protected" people remain in a situation of "dependency" instead of achieving the autonomy and independence

---

[1]If legislation permits the voluntary interruption of a pregnancy by a woman due to the fact that she learns that the unborn baby is a female, or a specific ethnicity or sexual orientation, many voices are raised to say the law is discriminatory or threatens the principle of equality. Nevertheless, the majority of laws regulating voluntary interruption of pregnancy include as a reason for abortion the act of discovering (or even suspecting) that the unborn child suffers from a physical, psychological or sensory deficiency. Although we recognise that the example is, as well as provocative, questionable, would the reader have the same doubts if the reason were that the unborn child was of a specific sex or ethnic group?

of one who exercises their rights and has them respected. In this way a large part of the exclusion of people with disabilities is incentivised by social assistance programmes, leading to imprisonment rather than liberation.

Social assistance systems are generally used to provide support to people with disabilities, or to compensate them for finding themselves outside the general flow of society. However, this requires daily life to be built only around people without disabilities, contributing more and more to the physical absence of those with disabilities in society in general.

(c) Finally we must also look at the "*physical barriers*" to the exercise of rights. For example, even when the right to vote is protected by law, in practice there are numerous factors which may make it impossible to exercise this right. These factors include the lack of accessible polling stations, accessible material (voting papers in Braille, for instance), transport to polling stations and the possibility of postal voting. Equally, the freedom of people with disabilities is limited by lack of accessibility. Most of us take for granted our ability to move around the physical environment without obstacles and to use normal channels of communication. We include all of this in our basic understanding of <freedom>. Nevertheless, although these abilities are not universal, ordinary services are not designed to include people who do not possess them. We must include accessible transport, the physical environment and accessible channels of communication as factors which damage the value of freedom to the extent that their existence for people with disabilities becomes doubtful.

Also, if the situation produced by these barriers can be generally applied to all people with disabilities, it is obvious that people with disabilities due to psychological deficiencies or mental illness constitute the most discriminated against group, as well as the most vulnerable. Beyond the discrimination practiced by society, we find many examples of the special discrimination people with disabilities suffer at the hands of the government. In many countries they are denied the right to education and learning, or they are refused entry to the country (even as tourists).

The most serious aspect of this is that many of these forms of discrimination are not promoted only by governments or society, but more importantly by the family environment, in this most private sphere where parents in their desire to "protect", "incapacitate". Sadly, legislation usually allows for the presumption, not always correct, that parents always know what is best for their child.

In conclusion, the legal structures created to promote private and public freedoms have not been applied or have been applied less rigorously in the case of people with disabilities. This has given rise to a category of people who, while depending on the public sphere for their survival, lack access to political power and influence. These people are denied full participation in public politics and full control over their individual destinies. They remain outside of the general flow of civil society. This invisibility or lack of presence, serves to reinforce stereotypical ideas about the incapacities of people with disabilities and feeds the lack of respect for them as holders of rights on an equal basis with others.

Despite these observations, when looking at the international law of human rights, some things stand out:

(a) The lack, until August 2006 of international treaties within the framework of the United Nations which refer to the rights of this group of people and,
(b) The fact that general treaties (The Bill of Human Rights) or those relating to other categories of vulnerable people (children, women, migrants . . .) include scant reference to people with disabilities.

The question that then arises is, why?

In my opinion, behind all of these considerations lies a double debate. First, if these people are *equal* or not to others (or if you prefer: if they are as "human" as others) and secondly and consequently, if they should be the *object* of regulation or *subject* to regulation.

This debate has raged over the course of the second half of the twentieth century. The starting point was a model characterised by social marginalisation; people with disabilities were treated as subjects for assistance and tutoring and as dependent and unproductive beings.

This model was superseded in the 1950s by the rehabilitation model, in which the individual was the protagonist who required recuperative medical treatment. The aim was to prevent disability or to rehabilitate the disabled person in order to make them "normal". If this were not possible the "discapacitated", "disabled", "deficient", "retarded" person was the object of compassion and charity and had to be "protected".

However, from the 1970s, there began to be a change of perspective in which charity stopped being the criteria applied to the analysis of disability and was replaced by human rights.

The origin of this change can be situated in the "*independent living*" movement which arose in the United States as a by-product of the Vietnam war. As opposed to previous wars, the American troops returned defeated. More than 50,000 American soldiers died on the South East Asian peninsula and several thousand were affected physically, mentally or sensorally. Their disability was compounded by the stigma of being "losers" in a society defined by competence.

Their limited social esteem had to be compensated for by a fight for their demands to be met and their rights fulfilled. The films "Coming Home" and "Born on the Fourth of July" are particularly illustrative of this period of the history of this country, which was also marked by the African-American population's fight for their civil rights and the rise of feminism.

This new perspective called for the disabled to be considered as subjects not as objects. The aim was to abandon the medical model which was centred on the medical aspects of disability and placed the "problem of disability" with the person, replacing it with a model of human rights which considers the individual as the centre of all decisions affecting them and removes the "problem" of disability from the person, placing it onto society. In this social model, the "problem" of disability derives, not from the conditions of the person, but from the lack of understanding by the State and civil society towards the difference that this disability represents, about what needs to be done to tackle the way in which the various economic and social processes affect or do not affect the difference implied by disability. It is the State which has the responsibility to tackle these socially created obstacles, with the aim of guaranteeing full respect for the dignity and equality of the rights of all persons. From this point the debate on the rights of people with disabilities became a wider debate on the *place that difference occupies in society and the role of the state in the integration of this difference.*

The ultimate objective of this social perspective of human rights is to construct societies which are authentically integrated, societies which value difference and respect the dignity and equality of all humans, regardless of their dissimilarities.

3.2 The Evolution of International Law According to the Social Evolution

The evolution of international law, especially within the framework of the United Nations is a reflection of these debates and this evolution.

Thus the first measures can be quantified as "benefit measures".[2] In the early years we do not find any international treaties on human rights that refer to the rights of people with disabilities.[3] It is true however that at this early stage we can find some unconventional texts, with a programmatic but not binding character, where reference is made, however indirectly, to disability.[4]

From the beginning of the 1970s, in response to the previously mentioned social movements, the situation begins to change with the adoption of various texts, which despite being only legal recommendations with often lamentable terminology, nevertheless provide the start of a human rights perspective in international law texts.[5]

---

[2]The first specific instrument of the United Nations in the sphere of disability was the Resolution on "*Social Rehabilitation of the Physically Disabled*" approved by the Economic and Social Council in 1950. This originated from a study by the Social Commission, a subsidiary body of the Economic and Social Council during its sixth session, with two reports relating to physical disabilities ("The social rehabilitation of the handicapped" and "The social rehabilitation of the blind") as well as another produced by the *International Program for the Wellbeing of the Blind.*

[3]Neither the Universal Declaration of Human Rights, nor the International Covenant on Civil and Political Rights, nor the International Covenant on Economic, Social and Cultural Rights, nor the first treaties on specific groups include any willingness to make direct reference to people with disabilities and their especially disadvantaged position.

[4]This is the case of the Declaration on the Rights of the Child from 1959 where one principle (of the ten set out) is devoted to children with disabilities (AG/RES/1386 (XIV) of 20 November 1959): or the Declaration on Social Progress and Development of 1969 (AG/RES/2542 (XXIV) of 11 December 1969).

[5]First in 1971 was the passing of the *Declaration on the Rights of Mentally Retarded Persons* (AG/RES/2856 (XXVI) of 20 December 1971) which became the first United Nations text to recognise specific rights for disabled people. Secondly in 1975 was the *Declaration on the Rights*

Finally, after the failed attempt to adopt a treaty, the most important document concerning the rights of the disabled within the framework of the United Nations up to the advent of the 2006 Convention, was approved: the *Rules on the Equalization of Opportunities for Persons with Disabilities* in 1993.[6]

Now in the twenty-first century, thanks to a proposal presented by Mexico, the following invitation to the General Assembly was included in the Action Programme adopted by the Durban 2001 World Conference Against Racism, Racial Discrimination, Xenophobia and Related Intolerance[7]:

> (The Conference) Invites the United Nations General Assembly to consider elaborating an integral and comprehensive international convention to protect and promote the rights and dignity of disabled people, including, especially, provisions that address the discriminatory practices and treatment affecting them[8]

In accordance with this invitation, again at Mexico's urging, the General Assembly approved resolution 56/168 on 19 December 2001, entitled *International Convention to promote and protect the rights and dignity of persons with disabilities*, which decides "to establish an Ad Hoc Committee, open to the participation of all member States and observers of the United Nations, to consider proposals for a comprehensive and integral international convention to promote and protect the rights and dignity of persons with disabilities, based on the holistic approach in the work done in the fields of social development, human rights and non-discrimination and taking into account the recommendations of the Commission on Human Rights and the Commission for Social Development".

On August 28, 2006, the aforementioned Special Committee approved the International Convention project on the rights of persons with disabilities and its Enforcing Protocol. The project became the first great human rights treaty of the twenty-first century to be adopted by the United Nations.

---

of Disabled Persons (AG/RES/3447 (XXX) of 09 December 1975). – Thirdly in December 1976 the General Assembly declared 1981 *International Year of Disabled Persons* with the aim of promoting wider integration and participation of these people in society. Fourthly, in December of 1982, the General Assembly adopted the *World Programme of Action Concerning Disabled Persons* (AG/RES/37/52 of 03 December 1982). This *Programme* is a declaration of principles and directives for national and international action in favour of people with disabilities. In the same resolution that approved the World Programme of Action for Disabled Persons, the General Assembly declared 1983–1992 as *Decade of Disabled Persons* with the aim of establishing a timeframe to allow the development and execution of activities included in the *World Programme of Action* and celebrate an International Convention directed at the specific protection of people with disabilities. The end of the 1980s saw the approval of the *Tallin Guidelines for Action on the Developments of Human Resources in the Field of Disability* and their incorporation by the General Assembly into an annex of resolution 44/70 of 08 December 1989. Before the end of the Decade of Disabled Persons, in 1991 the General Assembly adopted the *Principles for the Protection of Persons with Mental Illness and the Improvement of Mental Healthcare* (AG/RES/46/119 of 17 December 1991).

[6] AG/RES/48/96 of 20 December 1993.

[7] *World Conference against Racism, Racial Discrimination, Xenophobia and Related Intolerance,* Durban (South Africa), 31 August to 07 September 2001 (A/CONF.189/5).

[8] Ibid., Cap. I.

3.3 Redefining Disability

In accordance with article 1 of the Convention, the aim is to "promote, protect and fulfil the full and equal enjoyment of all human rights and fundamental freedoms of persons with disabilities in order to fully integrate them in their society and enable them to contribute in all its aspects with dignity." Logically the fundamental question that this raises is what one is to understand by disability. The second paragraph of the article makes it clear that this concept encloses "persons with disabilities include those who have long-term physical, mental, intellectual or sensory impairments which in interaction with various barriers may hinder their full and effective participation in society on an equal basis with others."

From this definition we can deduce that the concept of disability from which the Convention developed is based on the social model of the same; that it is not deficiency itself that limits participation in society, but interaction with different barriers. Such an *avant la lettre* concept of disability is not the result of chance, but rather of a long process.

The first attempt to define disability dates from 1980 when the World Health Organization published the International Classification of Impairments, Disabilities and Handicaps, ICIDH,[9] which made a distinction between *impairments, disabilities and handicaps.* According to this classification, *disability* was to be understood as "any restriction or lack (resulting from an impairment) of ability to perform an activity in the manner or within the range considered normal for a human being".[10] The practical application of ICIDH was immediate, and these were the definitions used in the World Action Programme. Nevertheless, this classification, born out of a desire to unify criteria on an international level, became outdated with the extension of the social model of disability.

In 2001, the WHO published a new version of the ICIDH: The ICIDH-2 which has come to be known as the International Classification of Functioning, Disability and Health (ICF).[11] This new version aimed to provide a new focus for the concept of disability and begun the search for an integrated vision, at once more complex and *interactional* and with a clear universal vocation. For the WHO, disability stops being something in itself and becomes the result of numerous variables. According to the ICF, disability is a concept which "includes impairments, limitations in

---

[9]As agreed in 1976 by the 29th World Health Assembly, in resolution 29.35. ICIDH or ICIDH-1 is the acronym for the first classification produced by the World Health Organization. Usually written as ICIDH in English.

[10]In the classification the definition of impairment is described as "any loss or abnormality of psychological, physiological or anatomical structure or function" and the definition of handicap States that a handicap is a "disadvantage for a given individual, resulting from an impairment or a disability, that limits or prevents the fulfilment of a role that is normal (depending on age, sex, and social and cultural factors) for that individual".

[11]Approved by the World Health Assembly in resolution 54.21 of 22 May 2001.

activity or restrictions on participation". Thus, disability is articulated through a dynamic interaction between personal conditions, conditions of health and the physical and social environment of each person.

Nevertheless the technical complexity of the ICF has made its use in legal and social science very difficult. An attempt was made to rectify this during the talks and negotiations prior to and during the conference. It was very difficult to reach an agreement on the concept of "persons with disabilities". The reasons for this were not only the diverse range of criteria, but also the desire of various States to not include a concept so restrictive as to leave many people out of the scope of the agreement, or so wide as to make its fulfilment by those States impossible. As STEIN points out "to secure broad support in the General Assembly, several of the Working Group members believed that these definitions ought purposely to be left vague so that States parties could interpret them according to their own legal and social cultures".[12] An agreement was finally reached on a concept that, as we have seen, as well as being sufficiently wide ranging, reflects the social model on which the recognition of rights is based. Nevertheless, this presents a serious danger. A vague definition of disability which provides each State with the legal authority to determine which suppositions will or will not be included may generate resolutions in which the same person is or is not subject to the application of the Convention depending in which state they find themselves.

3.4 The Principles of the Convention, Especially Accessibility and Nondiscrimination

Before analysing the concrete rules that regulate the specific rights protected by the Convention, it is worth looking at the general principles that inform them. The first observation to make is that these principles are the concrete translation of the general principles that inspire respect for and protection of human rights for all vulnerable groups and which were brilliantly outlined by the Inter-American Court of Human Rights in its well-known Advisory Opinion on the judicial condition and rights of undocumented immigrants.[13] In this, the Court has stated that: "the principle of equality before the law, equal protection before the law and non-discrimination belongs to *jus cogens*, because the whole legal structure of national and international public order rests on it and it is a fundamental principle that permeates all laws. Nowadays, no legal act that is in conflict with this fundamental principle is acceptable, and discriminatory treatment of any person, owing to gender, race, colour, language, religion or belief, political or other opinion, national, ethnic

---

[12] STEIN, M., "Disability Human Rights," *California Law Review* 95, no. 1 (2007).

[13] Inter-American Court of Human Rights: advisory opinion OC-18/03 of 17 September 2003 requested by the United Mexican States on *Juridical Condition and Rights of Undocumented Migrants.*

or social origin, nationality, age, economic situation, property, civil status, birth or any other status is acceptable".[14]

The same Court states that "this general obligation to respect and guarantee human rights, without any discrimination and on an equal footing, has various consequences and effects that are defined in specific obligations":

- First, "in compliance with this obligation, States must abstain from carrying out any action that, in any way, directly or indirectly, is aimed at creating situations of de jure or de facto discrimination".[15]
- Secondly, "in addition, States are obliged to take affirmative action to reverse or change discriminatory situations that exist in their societies to the detriment of a specific group of persons".[16]

The realisation of these obligations within the concrete framework of persons with disabilities, as established by the United Nations Convention, consists on the one hand, of the prohibition of all types of "discrimination for reasons of disability" and on the other hand the precaution of establishing "reasonable accommodation" so that society may have a "universal and inclusive design".

These concepts are developed in article 2 of the 2006 Convention, which defines "discrimination on the basis of disability" as "any distinction, exclusion or restriction on the basis of disability which has the purpose or effect of impairing or nullifying the recognition, enjoyment or exercise, on an equal basis with others, of all human rights and fundamental freedoms in the political, economic, social, cultural, civil or any other field. It includes all forms of discrimination, including denial of reasonable accommodation".

The same article defines "reasonable accommodation" as "necessary and appropriate modification and adjustments not imposing a disproportionate or undue burden, where needed in a particular case, to ensure to persons with disabilities the enjoyment or exercise on an equal basis with others of all human rights and fundamental freedoms".

Without doubt, achieving the non discrimination of these persons implies the need for the State to adopt measures to correct the situation of inequality from which they start. In this respect the Convention states that "specific measures which are necessary to accelerate or achieve de facto equality of persons with disabilities shall not be considered discrimination under the terms of the present Convention" (article 5, par. 4).

---

[14]Ibid., par. 101. In the same paragraph the Court continues: "This principle (equality and non-discrimination) forms part of general international law. At the existing stage of the development of international law, the fundamental principle of equality and non-discrimination has entered the realm of *jus cogens*".

[15]*Ibidem, par.* 103. In the same paragraph the Court continues: "This translates, for example, into the prohibition to enact laws, in the broadest sense, formulate civil, administrative or any other measures, or encourage acts or practices of their officials, in implementation or interpretation of the law that discriminates against a specific group of persons because of their race, gender, colour or other reasons".

[16]*Ibidem, par.* 104.

All this should give rise to a society where these measures are no longer necessary; a society of "universal design" and "inclusive design" defined in the convention as "the design of products, environments, programmes and services to be usable by all people, to the greatest extent possible, without the need for adaptation or specialized design".[17] Naturally, "universal design and inclusive design shall not exclude assistive devices for particular groups of persons with disabilities where this is needed".[18]

In other words, the two main principles that inspire this Convention are, in the first place, equality and non-discrimination, and secondly, accessibility.

In my opinion, it is in this sense that we must contemplate the general principles of the Convention stated in article 3.[19]

Evidently the majority of the principles recognised in this resolution are generally accepted principles and recognised by the International Community as a whole. For example, respect for inherent dignity, individual autonomy, including independence and the freedom to take decisions, non discrimination, equality of opportunity; or equality of the sexes.

But, this resolution also includes a series of principles, which due to their specific nature should be highlighted; the principle of "full and *effective* participation and *inclusion* in society"[20]; the principle of "respect for difference and acceptance of persons with disabilities as part of human diversity and humanity"[21]; the principle of "accessibility"[22]; and the principle of "respect for the evolving capacities of children with disabilities and respect for the right of children with disabilities to preserve their identities".[23]

The first of these principles – "full and effective participation and inclusion in society"- constitutes an important addition to the 2006 Convention, as it paves the

---

[17] Article 2 par. 5 of the Convention.

[18] *Ibidem.*

[19] The general principles of the Convention are stated in article 3 and are:

(a) Respect for inherent dignity, individual autonomy including the freedom to make one's own choices, and independence of persons;
(b) Non-discrimination
(c) Full and effective participation and inclusion in society
(d) Respect for difference and acceptance of persons with disabilities as part of human diversity and humanity
(e) Equality of opportunity
(f) Accessibility;
(g) Equality between men and women
(h) Respect for the evolving capacities of children with disabilities and respect for the right of children with disabilities to preserve their identities

[20] Article 3 section (c). Emphasis added.

[21] *Ibidem*, section (d).

[22] *Ibidem*, section (f).

[23] *Ibidem*, section (h).

way for the adoption of regulations of positive law and effective measures on the part of States so that people with disabilities enjoy the rights that inform this principle and that will lead to the effective integration of all people to the inclusive society of which the Convention speaks. These obligations on States are enforceable by the citizens who are direct beneficiaries of them: the people who face these barriers which "exclude" them and prevent them from participating.

In this respect, said principle must be related to the principle of "accessibility" which implies the elimination of all barriers preventing the exercising of rights. It is true that there is an article in the Convention, (number 9) which regulates the right of accessibility. However the principle of accessibility goes further than this. In accordance with article 9 of the Convention, in order for people with disabilities to be able to live independently and participate fully in all aspects of life, the participating States will adopt appropriate measures to ensure access to people with disabilities, under the same conditions as others, to the physical environment, transport, information and communication, including information technology and communication systems and to other public services and installations, both in rural and urban areas.[24] But, the "principle of accessibility" is not limited to access to installations and technology, that is to say, to the elimination of "physical barriers", but it also includes "access to the enjoyment of all rights", in other words it leads to the creation of a society accessible to all, in which all barriers disappear. Therefore,

[24]In accordance with the text of article 9: "These measures, which shall include the identification and elimination of obstacles and barriers to accessibility, shall apply to, inter alia:

1. Buildings, roads, transportation and other indoor and outdoor facilities, including schools, housing, medical facilities and workplaces;
2. Information, communications and other services, including electronic services and emergency services.

Also, in accordance with section 2 of article 9: States Parties shall also take appropriate measures to:

1. Develop, promulgate and monitor the implementation of minimum standards and guidelines for the accessibility of facilities and services open or provided to the public;
2. Ensure that private entities that offer facilities and services which are open or provided to the public take into account all aspects of accessibility for persons with disabilities;
3. Provide training for stakeholders on accessibility issues facing persons with disabilities;
4. Provide in buildings and other facilities open to the public signage in Braille and in easy to read and understand forms:
5. Provide forms of live assistance and intermediaries, including guides, readers and professional sign language interpreters, to facilitate accessibility to buildings and other facilities open to the public;
6. Promote other appropriate forms of assistance and support to persons with disabilities to ensure their access to information;
7. Promote access for persons with disabilities to new information and communications technologies and systems, including the Internet;
8. Promote the design, development, production and distribution of accessible information and communications technologies and systems at an early stage, so that these technologies and systems become accessible at minimum cost".

for example, some barriers must disappear, such as those that impede the exercising of political rights, particularly the right to active and passive suffrage (complete prohibition or making access to voting impossible); rights to family (prohibiting marriage, adoption, etc.) and privacy; rights of defence (denying the deaf and blind to a legal defence due to a lack of interpreters, considering an illiterate person incapable of defending themselves); without mentioning the problems relating to the right to life in cases of extermination of newborns (above all in countries practising demographic controls), selective withholding of medical treatment, legislation on euthanasia or the controversial abortion for reasons of euthanasia.

These two principles – "full and *effective* participation and *inclusion* in society" and "accessibility"-, relating wholly to the creation of an integrated and inclusive society, are also reflected not only in the mentioned principles but also in a series of rights recognised by the Convention. It is article 19 that recognises the right to live independently and be *included* in the community, anticipating the obligation of States to adopt effective and appropriate measures to facilitate full enjoyment by persons with disabilities to live independently; and article 24 which recognises the right to an *inclusive* education system at all levels and lifelong learning. In order to make this right effective, it stipulates that States will ensure that persons with disabilities are not excluded from the general education system on the basis of disability, and that they can access an *inclusive* primary and secondary education on an equal basis with others, making "reasonable accommodations" depending on individual necessities. Articles 26 and 27 are also consequences of these principles. Article 26 recognises the rights to habilitation and rehabilitation to achieve and maintain maximum independence, full physical, mental, social and vocational ability, and full *inclusion* and *participation* in all aspects of life, obliging States to adopt appropriate measures to ensure this. Finally, article 27 recognises the right to earn a living through a freely chosen profession in an open market or work environment which are *inclusive of* and *accessible to* persons with disability.

The other two principles highlighted have a different sense. They are the principles of "respect for difference and acceptance of persons with disabilities as part of human diversity and humanity" and "respect for the evolving capacities of children with disabilities and respect for the right of children with disabilities to preserve their identities". These are two principles which permeate the entire Convention and relate to the demand for respect for the inherent difference among all human beings.

In this respect, it can be said that the 2006 Convention is not merely a treaty relating to the rights of people with disabilities, but rather a treaty which enriches the cultural tradition of human rights, providing a new paradigm: we are all different in some way from other members of society. Society should therefore recognise these differences as enriching, accepting and integrating them so they do not impinge on "the recognition of the inherent dignity and worth and the equal and inalienable rights of all members of the human family".[25]

---

[25]First paragraph of the preamble to the 2006 Convention.

3.5 The Nucleus of the Convention: Rights

One of the first debates to arise when the contents of the first draft were being negotiated was how to determine the model or paradigm that was to be adopted. On the one hand, there were those in favour of adopting a comprehensive or "holistic" approach to disability, which entailed including the different categories of human rights adapted to the specific requirements of people with disabilities to exercise said rights. The other main focus was that of non-discrimination, according to which, rather than merely recognising specific rights, the Convention would have to include measures to allow people with disabilities the possibility of exercising rights already recognised by other bodies. Finally, the model adopted by the Convention was one of a "mixed" focus, which represented an important first in the International law of Human Rights. It is important to bear in mind that as human beings, people with disabilities have the same recognised rights as any other person. Therefore, the Convention recognises a catalogue of rights, which without being new rights, are adapted to the specifics of people with disabilities. This adaptation assumes that "none of these rights is formulated in a reiterative form in comparison with other human rights treaties, but rather refer to the specific circumstances of people with disabilities".[26] And the principles of non-discrimination and accessibility interact with each of these rights, ensuring their effective exercise by each person with a disability.

If we look at the classic division of human rights into stages, among the wide range of human rights included in the Convention we find civil, political, economic, social and cultural rights. However, to better illustrate the specific rights recognised, the classification proposed by GERARD QUINN is particularly useful in its description of the general principles proclaimed in article 3.[27] In this way we can differentiate between rights relating to dignity, freedom, rights of autonomy and independence, rights of inclusion and participation and solidarity rights. Nevertheless, these groups are profoundly interconnected and it is not always easy to draw lines between them.

The rights to dignity include the right to life (article 10), protection in **situations of risk and humanitarian emergencies** (article 11), protection against torture and other cruel, inhumane or degrading treatments or punishments (article 15), protection against exploitation, violence and abuse (article 16) and protection of the integrity of the person (article 17). These rights, as essential as they are basic, acquire a special dimension for those with disabilities. It is not unusual to hear of cases of abortion due to the fact that the foetus had a disability, cases of illegal

---

[26] CARDONA LLORENS, J. y SANJOSÉ GIL, A., 2005. *La promoción y protección de los Derechos Humanos de las personas con discapacidad: en el camino de su reconocimiento universal*, in COURTIS C., HAUSER D. y RODRÍGUEZ HUERTA G. (comp.), *Protección Internacional de Derechos Humanos. Nuevos Desafíos*, (México: Instituto Tecnologico Autonomo de Mexico, 2005), pp. 147–175.

[27] QUINN, G., *The UN Human Rights of Persons with Disabilities Treaty: A Blueprint for Disability Law & Policy Research and Reform*, 2006.

imprisonment in institutions, medical treatment without full informed consent, etc. The new rights of protection in situations of risk, including armed conflicts, is positive as wars and conflicts are a devastating source of disability, affecting women and children particularly.[28]

The rights to freedom are represented by the right to liberty and security of the person (article 14), liberty of movement and nationality (article 18), personal mobility (article 20) and freedom of expression and opinion, and access to information (article 21). The right to personal mobility stands out from among these rights because of its specific relation to the context of disability. It involves facilitating the mobility of people with disabilities (in the manner and the time of their choice) and access for people with disabilities to quality mobility aids, devices, assistive technologies and forms of life assistance and intermediaries. The costs for this are to be affordable, but what is meant to be "affordable" has yet to be defined.

The rights of autonomy and independence include the right to equal recognition before the law (article 12), respect for privacy (article 22) as well as respect for home and family (article 23), habilitation and rehabilitation (article 26) and the right to live independently and be included in the community (article 19). This last right however, is a halfway point between this group of rights and the group relating to the rights of inclusion and participation. Only through independent living with serious guarantees (to the maximum degree possible) can a person aspire to form part of the community around them.

Another element to highlight is that rehabilitation is not the end, as it was with the medical model of disability. It is now -in accordance with article 26- another right; an element by which people with disability can achieve and maintain maximum independence and physical, mental, social and vocational abilities and full inclusion and participation in all aspects of life.

The rights of inclusion and participation cover the rights to accessibility (article 9), access to justice (article 13), participation in political and public life (article 29) and participation in cultural life, recreational activities, leisure activities and sport (article 30). In this sense, the States must guarantee people with disabilities the opportunity to participate effectively and completely on an equal basis with others.

Lastly, QUINN speaks of solidarity rights in reference to the right to education (article 24), the right to health (article 25), the right to work and employment (article 27) and the right to an adequate standard of living and social protection (article 28). Although solidarity is not specifically included as a principle in the Convention GERARD QUINN and THERESIA DEGENER previously referred to it, citing it as of essential importance in the human rights discourse.[29] For Quinn and Degener, solidarity, based on the recognition of the existence of the links and mutual obligations between people, consists of preparing people for participation in all social processes.[30]

[28] See UNHCR, *Disabled Persons, Victims of Armed Conflicts and Civil Unrest*, Eighth inter-agency meeting on the United Nations Decade of Disabled Persons, 1983–1992, Geneva, 1990.

[29] QUINN, G. y DEGENER, T., *Derechos humanos y discapacidad*, cit., pp. 11 y ss.

[30] *Ibidem*, p. 15.

The right to work is regulated exhaustively. Work is, for the vast majority of people, the main source of subsistence and allows them to exercise many other rights. It facilitates economic independence and constitutes an essential part of the human condition. Discrimination in the work environment because of disability therefore constitutes one of the gravest and most common forms of discrimination. The unemployment rate for people with disabilities is, in most of the world, two or three times higher than for able bodied people. In addition, the positions usually offered to people with disabilities are low grade positions, badly paid, with little responsibility and in general way below the disabled person's capabilities. This is why the right to work of people with disabilities occupies such a central place in the Convention.

3.6 Enforcing Rights

Of these rights recognised by the Convention, are they all directly applicable, or on the contrary, do some of them depend on the economic development of the participating States, and could they be considered "social policy directives" rather than real rights?

Many of the articles of the Convention refer to economic, social and cultural rights, which, as we know, is the arena where this debate usually takes place.

In the preparatory work prior to the Convention, the States maintained a reductionist position on the rights discussed, giving the economic, social and cultural rights a progressive character, dependent on development.

When debated by the Committee, a rigid demarcation between civil and political rights and economic, social and cultural rights was felt to be unhelpful. Therefore, on the one hand attention was drawn to the many articles of the project with a mixed character, and on the other an attempt was made to avoid creating a "hierarchy" of rights within the Convention. However the truth is that a "hierarchy" has developed and the distinction between one type of right and another is evident throughout the Convention.

One of the clearest examples can be seen in article 4 which relates to the obligations of States. Although there was general agreement in affirming that the States' parties were obliged to adopt all appropriate measures to "ensure the full realization of all the rights established in the Convention", this general agreement also stated that "with regard to economic, social and cultural rights, each state party undertakes to take measures to the maximum of its available resources and, where needed, within the framework of international cooperation".

Finally, after a promising first paragraph which states:

> 1. "States parties undertake to ensure and promote the full realization of all human rights and fundamental freedoms for all persons with disabilities without discrimination of any kind on the basis of disability. To this end, States parties undertake to . . ."[31]

[31]The undertakings are:

The second paragraph is less hopeful, stating that:

> 2. "With regard to economic, social and cultural rights, each state party undertakes to take measures to the maximum of its available resources and, where needed, within the framework of international cooperation, with a view to achieving progressively the full realization of these rights, without prejudice to those obligations contained in the present Convention that are immediately applicable according to international law."

This very same question was raised several times during negotiations on the specific rights. Therefore, when debating article 15 for example, which relates to the right to live in the community and the right to live independently on an equal basis with others, it was noted that the sections relating to support services and installations needed to enable these rights referred to economic, social and cultural rights and *were therefore susceptible to progressive realization.* It was pointed out that "although article 4 included the concept of progressive realization, this does not prevent the use of the term progressive realization in article 15 or in other articles with mixed or ambiguous resolutions".[32]

Having accepted this formula, the right to live in the community and the right to live independently on an equal basis with others became dependent on the existence of resources to adopt the support measures, which implies the ultimate denial of the right, although it forms part of the category of civil and political rights.

(a) adopt all appropriate legislative, administrative and other measures for the implementation of the rights recognized in the present Convention;
(b) take all appropriate measures, including legislation, to modify or abolish existing laws, regulations, customs and practices that constitute discrimination against persons with disabilities;
(c) take into account the protection and promotion of the human rights of persons with disabilities in all policies and programmes;
(d) refrain from engaging in any act or practice that is inconsistent with the present Convention and to ensure that public authorities and institutions act in conformity with the present Convention;
(e) take all appropriate measures to eliminate discrimination on the basis of disability by any person, organization or private enterprise;
(f) undertake or promote research and development of:

    i. universally designed goods, services, equipment and facilities, as defined in article 2 of the present Convention, which should require the minimum possible adaptation and the least cost to meet the specific needs of a person with disabilities, to promote their availability and use, and to promote universal design in the development of standards and guidelines;
    ii. new technologies, including information and communications technologies, mobility aids, devices and assistive technologies, suitable for persons with disabilities, giving priority to technologies at an affordable cost;

(g) provide accessible information to persons with disabilities about mobility aids, devices and assistive technologies, including new technologies, as well as other forms of assistance, support services and facilities;
(h) promote the training of professionals and staff working with persons with disabilities in the rights recognized in this Convention so as to better provide the assistance and services guaranteed by those rights.

[32] See comment on article 15, in A/60/266, par. 21.

Analysis of the debates in other articles, such as that in article 7 relating to non consideration as discrimination against the measures designed to accelerate equality show us that this question has been raised in a number of States.

Nevertheless we should congratulate ourselves on the disappearance of any reference to "progressively achieving realization" or "maximum available resources" in the definitive Convention text. All these proposals from the previous drafts disappeared.

These "disappearances" should be welcomed optimistically. As the Committee has pointed out, the majority of resolutions contained in the project are mixed, in the sense that the majority of articles referring exclusively to the recognition of civil or political rights also contain obligations for the States to adopt measures to compensate for the situation of inequality and greater vulnerability of people with disabilities. If these measures are not adopted, exercise of the rights would be impossible. However, these measures are seen by States as falling within the arena of economic, social and cultural rights, that is to say, social policy measures.

This meant that the final result was that the dependence on economic development for the adoption of these measures creates dependence on economic development for the exercise not only of economic, social and cultural rights (which is, in itself debatable), but also the exercise of civil and political rights.

The result is discriminatory and does not leave people with disabilities "on an equal basis" with others.

Nevertheless, with the disappearance of the references to progressive nature and available resources from the resolutions, we can state that all the measures which must be taken by the States in order to fully realize civil and political rights – measures which have previously been seen as belonging to social policy (that is to say, outside of the sphere of right), or at most as a social right – become as obligatory as civil or political rights, the exercise of which allows a level of equality.

However, it is important to highlight paragraph 2 of article 4 which places the exercising of economic, cultural and social rights within the arena of the progressive realization and maximum available resources.

Despite this, I would like to point out two things. First, it is not the economic, social and cultural rights which refer to progressive realization, but the measures necessary to achieve their full realization. It is true that recognising the right but not the immediate requirement for the implementation of the measures necessary for its full realization by people with disability seems ineffective. However, in my opinion, the act of currently recognising the right allows juridical control of the concept of "to the maximum of its available resources" as stated in the same article.

Secondly, it is worth noting the last sentence of section 2 of article 4, which States that the progressive and limited nature of the resources should be understood "without prejudice to those obligations contained in the present Convention that are immediately applicable according to international law". In this respect I feel the reflections by the Committee on Economic, Social and Cultural Rights in its

General Comment number 3 are particularly relevant in relation to the "nature of States parties obligations" relating in similar format to that used in paragraph 1 of article 2 the Committee's opinion on the *progressive nature*:

> "... although the full realization of appropriate measures may be achieved gradually, the measures tending to achieve this objective should be adopted within a reasonably short time after the covenant becomes binding on participating States. It thus imposes an obligation to move as expeditiously and effectively as possible towards the obligations recognised in the covenant ... ", "... the fact that realization over time, or in other words progressively, is foreseen under the Covenant should not be misinterpreted as depriving the obligation of all meaningful content It is on the one hand a necessary flexibility device, reflecting the realities of the real world and the difficulties involved for any country in ensuring full realization of economic, social and cultural rights On the other hand, the phrase must be read in the light of the overall objective, indeed the raison d'être, of the Covenant which is to establish clear obligations for States parties in respect of the full realization of the rights in question. It thus imposes an obligation to move as expeditiously and effectively as possible towards that goal. Moreover, any deliberately retrogressive measures in that regard would require the most careful consideration and would need to be fully justified by reference to the totality of the rights provided for in the Covenant and in the context of the full use of the maximum available resources ... "

And the Committee considers that:

> a minimum core obligation to ensure the satisfaction of, at the very least, minimum essential levels of each of the rights is incumbent upon every state party ... If the Covenant were to be read in such a way as not to establish such a minimum core obligation, it would be largely deprived of its raison d'être. By the same token, it must be noted that any assessment as to whether a state has discharged its minimum core obligation must also take account of resource constraints applying within the country concerned.

Secondly, regarding the limit of "*to the maximum of its available resources*" the Committee mirrors the phrase used in the covenant:

Article 2 (1) obligates each state party to take the necessary steps "to the maximum of its available resources". For any state party to attribute its lack of compliance with the minimum obligations to a lack of available resources, it must demonstrate that it has made all efforts to utilise the resources at its disposal in an attempt to satisfy these minimum obligations.

> The Committee wishes to emphasize, however, that even where the available resources are demonstrably inadequate, the obligation remains for a state party to strive to ensure the widest possible enjoyment of the relevant rights under the prevailing circumstances. Moreover, the obligations to monitor the extent of the realization, or more especially of the non-realization, of economic, social and cultural rights, and to devise strategies and programmes for their promotion, are not in any way eliminated as a result of resource constraints.

It is therefore clear that the rights exist and that States must take at least the minimum measures required for them to be exercised and where it is not possible to adopt any of the measures, they must demonstrate that they have used all available resources to implement this measure although it has not been ultimately possible to achieve it. And these obligations are directly enforceable.

3.7 The Role of the Convention in International Human Rights Law

I have tried, in the preceding pages to show how the Convention searches for a balance between the idea that the only thing to ensure is non-discrimination and the other theory that affirms the existence of specific rights for people with disabilities. For the Convention, people with disabilities have the same rights as others they do not have specific rights. However, the States are obliged to adopt all necessary measures to achieve a society which is authentically inclusive, a society which values difference and respects the dignity and equality of all human beings, despite their differences and in which everyone can exercise their rights equally. In order to do this, the States' parties must adopt policies which eliminate existing barriers.

These obligations on States are enforceable by citizens, especially by their direct beneficiaries, those people who encounter the barriers. But not only by those affected. All citizens, with or without disabilities have the right to an inclusive and integrated society, which respects the dignity and equality of all.

In conclusion, the right exists subject to the state adopting these measures, but not as a specific right for people with disabilities, but rather a right of all people within society.

In this respect, it can be said that the 2006 Convention is not merely a treaty relating to the rights of people with disabilities, but a treaty which enriches the cultural tradition of human rights, providing a new paradigm; we are all different in some way from other members of society, and society should recognise these differences as enriching, accepting and integrating them so they do not impinge on "the recognition of the inherent dignity and worth and the equal and inalienable rights of all members of the human family".[33]

The 2006 Convention thus completes the process that began with other treaties relating to groups that were "different" and therefore "vulnerable" -women, infants, the stateless, native and migrant workers- and whose aim has been to remove the obstacles that prevent them exercising their rights. In the "gradual move towards the determination of the subjects of rights, according to a specification which has been produced regarding the genre, regarding the different stages of life, bearing in mind the difference between the normal state and the exceptional state in the human condition" of which BOBBIO speaks in the framework of what defines the "specification process of human rights",[34] international human rights law States that these "differences" should not be relevant to the exercising of rights and that we all have the right to a society which accepts and respects difference, without affecting the respect for and protection of all rights for all people.

In this respect, the 2006 Convention represents the final step in the general recognition of a right to an inclusive and integrated society; a right we all share and contribute to, not only those who are "different". In the end it is the proclamation that "we are all different" and therefore equal before the law.

[33]First paragraph of the preamble to the 2006 Convention.

[34]*Vide supra.*

3.8 Conclusions

As I stated in the introduction, legal structures to promote private and public freedoms have traditionally been applied or have been applied with less rigour in the case of people with physical, psychological, mental or sensory deficiencies, meaning they have different abilities to those considered "normal". This has given rise to a category of people who, depending on the public sphere for their survival, have lacked access to political power and have not been able to influence it. These people have been denied full participation in public policy and full control over their individual destinies. They have remained outside of the general flow of civil society. This invisibility or lack of presence serves to reinforce the stereotypical ideas about the incapacities of people with different abilities and feeds the lack of respect for them as holders of rights on an equal basis with others.

In the end, behind all of these considerations there is a double debate: first, if these people are or not *equal* to others (or if you prefer: if they are as "human" as others) and secondly and consequently, if they should be the object of regulation or *subject* to regulation.

As I have tried to demonstrate in this article, the 2006 Convention ended this debate. With it, the medical model which is centred on the medical aspects of disability and places the "problem of disability" with the person is abandoned, and replaced for a model of human rights which considers the individual as the centre of all decisions affecting him and places the "problem" of disability away from the person and onto society. In this social model, the "problem" of disability derives, not from the conditions of the person, but from the lack of understanding from the state and civil society towards the difference that this different ability represents and what needs to be done to tackle the way in which the different economic and social processes affect or do not affect the different abilities derived from physical, psychological, mental or sensory deficiencies. And it is the State which has the responsibility for tackling these socially created obstacles with the aim of guaranteeing full respect for the dignity and equality of the rights of all persons. From this point the debate on the rights of people with diverse abilities has become wider debate regarding the place difference occupies in society and the role of the state in the integration of this difference.

The ultimate objective of this social perspective of human rights is to construct societies which are authentically integrated, societies which value difference and respect the dignity and equality of all humans, regardless of their differences.

This is the challenge of human rights in the twenty-first century: build a society which is authentically inclusive, a society which values difference and respects the dignity and equality of all human beings, despite their differences and in which everyone can exercise their rights equally. Until then it will be necessary to implement policies which eliminate the barriers created by a society which has not been constructed as it should have been.

Part II
Human Rights: Hard Threats

Chapter 4
Afghanistan: Why Has Violence Replaced Political Power?

Encarnación Fernández Ruiz-Gálvez

Abstract The hypothesis from which Chap. 4 stems, and which I will try to confirm through this article, is that in Afghanistan it has been the international factors (albeit interacting with internal dynamics) which have had a particular hand in intensifying the conflicts and the process of State decomposition. On top of the underlying and undeniable structural weakness and external vulnerability of the Afghan State (a matter addressed in Sects. 4.1 and 4.2), the international interventions from the Afghan-Soviet war to the 2001 invasion, the lack of interest shown by the international community in consolidating peace in Afghanistan during the 1990s, and especially the numerous errors made by the international community led by the EU in the past decade, and which began with the invasion at the end of 2001, have all contributed to the constant decline in the country conditions (a thesis developed in the central part of this work: Sects. 4.3, 4.4, 4.5 and 4.6). Among other things, these factors have generated an environment ripe for the slow, progressive substitution of politics for violence which is one of the major obstacles today facing any attempts to reconstruct the country. In this sense, Hannah Arendt's analysis of the distinction between political power and violence could serve as a basis for reflection about the possible (and doubtless difficult) ways out of the current situation in Afghanistan, through a recuperation of the political field, in the most noble sense of the term, as a way of overcoming violence through communication; a question which is addressed further in the final section of this work.

E. Fernández Ruiz-Gálvez (✉)
Philosophy of Law, Department of Law, Moral and Political Philosophy, Master Human Rights, Peace and Sustainable Development, University of Valencia, Valencia, Spain
e-mail: Encarnacion.Fernandez@uv.es

J. Ballesteros et al. (eds.), *Globalization and Human Rights*, Ius Gentium: Comparative Perspectives on Law and Justice 13, DOI 10.1007/978-94-007-4020-4_4,

4.1 Historical and Geographical Context. State Weakness and External Vulnerability

Afghanistan's luck is tempered firstly by its geographical position. "For few countries in the world is it more true that geography determines history, politics and the nature of a people".[1] In the case of Afghanistan, the importance of geopolitics is clear, as it has been throughout its long history, and as it continues to be.[2]

Afghanistan's territory, with no exit to the sea, occupies a strategic position between Iran and India, and between Central Asia, the South of Asia and the Arabian Sea. Afghanistan shares borders with three great regions of Asia: Central Asia, the Indian subcontinent and the Middle East. It has relations with all of these, yet is different from all of them. From this it has been said of Afghanistan that it is the "crossroads of Asia".

Historically, Afghanistan have occupied an exceptional role in the commercial and invasion routes.

The battlefield status, along with being the object of rivalry between the great powers and/or regional states, without Afghanistan ever actually mattering except for the strategic position it occupies, was a constant in its historical trajectory. The "Great Game" which developed in the nineteenth century between the Czarist and the British empires for the control of Central Asia, with Afghanistan as the principle stage, and the Afghan-Soviet war of the 1980s, were two manifestations of this. Later, since the 1990s, Afghanistan has been at the centre of the battle for the oil and gas lines of Central Asia as for control of the commercial routes. Regarding the American occupation of 2001, Marc W. Herold states that the idea was to avoid the presence of a hostile government in an area of great geopolitical importance, recreating the historical function of Afghanistan as buffer state. There was no authentic interest in either the population or in establishing a real government.[3]

Afghanistan was constituted as a separate and independent state in 1747, when Ahmad Shah Abdali,[4] founder of the Durrani dynasty and considered to be the father of the nation, managed to glue together a tribal confederation independent of both the Safavid monarchy in Persia and the Moghuls of India.[5] Nonetheless, for more

---

[1]RASHID, Ahmed, *Taliban: Islam, Oil and the New Great Game in Central Asia* (London: I.B. Tauris, 2002), p. 7.

[2]PIRIS, Alberto, "Prólogo" a HEROLD, Marc W, *Afganistán como un espacio vacío. El perfecto Estado neocolonial del S. XXI* (Madrid: Foca, 2007), p. 9.

[3]HEROLD, Marc W, *Afghanistan as an Empty Space: The Perfect Neo-Colonial State of the 21st Century*, revised and updated April 2006, *passim*. Available at www.grassrootspeace.org/m_herold_afghanistan_0406.pdf.

[4]The Abdalis were one of the two great rival groups of the Pashtun tribes- Ghilzais and Abdalis (later known as Durranis).

[5]"The disintegration of two mighty empires, the Saffavid in Iran and the Mughal in India, provided the opportunity for a great tribal confederation of Pashtuns in 1747" (HYMAN, Anthony, *Afghanistan Under Soviet Domination, 1964-83* (London: Macmillan Press, 1984), p. 4).

than a century, and until the time of the second Anglo-Afghan war (1878–1880), the internal institutions were very weak. For this reason, for many the origins of the modern state in Afghanistan lie in the kingdom of Amir Abdul Rahman (1880–1901), known as the "Iron Amir", who introduced an unprecedented bureaucratic state in Afghanistan, and forcefully subdued the various regions of the country and the many ethnic groups to a central authority.[6]

One could conclude that in effect, under Abdul Rahman Afghanistan continued to be a weak state, with a barely articulated national identity and a clear lack of political legitimacy of the state itself. National identity does not requires homogeneity, only a sense of community among the state components which cannot be imposed co-actively. Social cohesion cannot be achieved through imposition and repression. Social integration achieved through these measures is only superficial and always very precarious. Definitively, there was a lack of social adhesion to the State, as illustrated by the numerous ethnic revolts which occurred during Rahman's reign. The Emir's modus operandi was a clear manifestation of what Mason calls "the weak states paradox": "insofar as its institutional coherence and legitimacy diminish, there is frequent recurrence to military repression as an instrument to maintain the strength of the regime and internal control".[7] In keeping with this analysis, the states who govern through repression and coercion are not strong states, although they are often considered so; in fact they are regimes with powerful central governments, but with a very weak legitimating concept of the state, thus making them weak states. The Emir's repressive policies simply led to a state of the sort Rotberg refers to as "seemingly strong" weak states.[8] The government is strong, but the State is weak.

According to Guistozzi's interpretation, the country's social heterogeneity hampered initial attempts to establish and consolidate the State through national construction. As a result, Abdul Rahman adopted an "imperial" model of State construction, maintaining control through coercion and through divide and rule policies. The imperial model of governance, though relatively resistant to adversity (Abdul Rahman successfully faced up to numerous insurrections against his regime), suffered serious internal cracks, to the extent that if, for whatever reason, the central government fell into crisis, the entire state system could also enter into crisis and fall apart. Here lies the underlying structural weakness and cyclical crisis

[6]MALEY, William, *The Afghanistan Wars* (Gordonsville: Palgrave Macmillan, 2002), "The Afghan State", pp. 11–3; RASANAYAGAM, Angelo, *Afghanistan: A Modern History. Monarchy, Despotism or Democracy? The Problems of Governance in the Muslim Tradition*, pp. 11–13 (London/New York; I.B. Tauris, 2010); and RASHID, Ahmed, *Taliban: . . . , cit.*, p. 12.

[7]MASON, A., "La crisis de seguridad en Colombia: causas y consecuencias internacionales de un Estado en vía de fracaso," *Revista Colombia Internacional*, no. 49/50 (junio 2001).

[8]ROTBERG, Robert I., "The Failure and Collapse of Nation-States: Breakdown, Prevention and Repair," in *When States Fail: Causes and Consequences,* ed. ROTBERG, Robert I., p. 5 (Princeton/Oxford: Princeton University Press, 2004).

of the Afghan State, throughout whose history one finds periods of consolidation repeatedly followed by more periods of collapse.[9]

For over 200 years (from the founding of the kingdom of Afghanistan in 1747 until 1973 when the last king Zahir Shah was deposed, and until the Marxist coup d'etat in 1978, given that President Daud who governed until that point was a member of the royal family), Afghanistan had been ruled by one or another of the Durrani clans. Thus, its internal politics was dictated by the Pashtuns, who Jawad refers to as the "dominant minority", but the Pashtun intent to impose an ideal of national unity inevitably failed, given that its true objective was always to extend their power.[10]

The disputes and rivalries among the Ghilzais and the Durranis and among the Pashtuns and non-Pashtuns continued throughout these two centuries, "but the country was governable and sustainable, despite its extreme poverty".[11] This was likely due to the persistence of the Durrani dynasty, who despite the weakness of their power had served as a factor of political unity "that could sometimes transcend internal rivalries and divisions".[12]

Furthermore, traditionally, Islam has granted an element of unity to Afghan society,[13] and a certain legitimacy to the state, but without power being exercised by religious leaders or authorities. Beyond this, "inquisition based on adherence to certain arbitrarily designated orthodox beliefs has not been part of the past of the Afghans".[14] This began to change when the Mujahedin came to power in 1992 and most strongly with the later government of the Taliban.

In addition to its significant internal fissures, Afghanistan has always been exposed to considerable external vulnerability due to is location in a buffer zone: geographically, between imperial Britain and Czarist Russia in the nineteenth century and early twentieth century; and ideologically, between the United States and the USSR in the Cold War period.[15]

Throughout the nineteenth century the British and Czarist empires (the Lion and the Bear, as they were presented in caricatures by the British magazine *Punch*) competed for influence in Kabul and for control of Afghanistan. Both empires carried out a clandestine war full of intrigue, trickery, secret agents and on

---

[9]GIUSTOZZI, Antonio, "Afghanistan: The Patrimonial Trap and the Dream of Institution-Building," in *Persistent State Weakness in the Global Age,* ed. KOSTOVICOVA, Denisa and BOJICIC-DZELILOVIC, Vesna, pp. 69–71 (Farnham/Burlington: Ashgate, 2009).

[10]JAWAD, Nassim, *Afghanistan. A Nation of Minorities*, Minority Rights Group Report, London, 1992, pp. 7, 9 and 14.

[11]RASHID, Ahmed, *Descent into Chaos. The World's Most Unstable Region and the Threat to Global Security* (London: Penguin Books, 2008), p. 8.

[12]RASANAYAGAM, Angelo, *Afghanistan: A Modern History . . .*, cit., p. 11.

[13]DOLLOT, René, *L'Afghanistan: Histoire. Description. Moeurs et coutumes. Folklore. Fouille* (Paris: Payot, 1937), p. 47.

[14]GHANI, Ashraf, "Impact of Foreign Aid on Relation Between State and Society," *Journal of Afghan Affairs* (Peshawar, Pakistan) 5, no. 4 (1990).

[15]JAWAD, Nassim, *Afghanistan. A Nation of Minorities*, cit., p. 7.

occasion military pressure, known as the "Great Game".[16] The Anglo-Afghan wars, especially the two first ones (1839–1842 y 1878–1880), were part of the "Great Game". Rubin claims that at the end of the second one the Afghan State had been formally established as a buffer state under British protection.[17]

Afghanistan's borders were established at the end of the nineteenth century, based on agreements between the two empires, resulting in the division of the tribes and ethnic groups into different States. The course of the Amu-Darya was adopted, along general lines, as the border between Russia and Afghanistan (Anglo-Russian agreements of 1873 and 1883).

In this way, the Tajiks, Uzbeks and Turkmen were separated on either side of the Amu-Daria; both in Afghanistan and in the Russian territories of Central Asia which today make up the ex-Soviet republics of Tajikistan, Uzbekistan and Turkmenistan. The borders with India, imposed by the British, were established in 1893 based on a treaty which formalized the division of the Pashtun and the Baluchis on either side of the Durand Line (thus named for Sir Mortimer Durand). From 1893 to 1895 the Anglo-Russian Commission of Pamir set the northern borders of Afghanistan up to Chinese territory, granting Afghanistan the Wajan corridor (an area situated in the northeast of the country, separating British India- Kashmir and Chitral, from the Czarist empire- currently Tajikistan) against the wishes of the Emir. This was a clear demonstration of the function assigned to Afghanistan as a buffer state. All of this was ratified in the Anglo-Russian Convention of Saint Petersburg in 1907. As can be seen, Afghanistan's geographical space, its territory was "defined by external powers", "without Afghan participation".[18]

Neither the Emir Rahman nor any later Afghan governor- not even during the Taliban regime- accepted the Durand Line. This was, and continues to be, a non-existent and permeable border and a permanent factor of conflict in the Anglo-Afghan relations and later in relations with Pakistan. Not in vain do the Pashtun tribes live mounted on horses on both sides of the border, arriving from the Pakistani side, through independent tribal territories, up to the Indus, which would be the true ethnic frontier of Afghanistan.

---

[16]RASHID, Ahmed, *Taliban:* ..., cit., p. 11.

[17]RUBIN, Barnett R., "Afghanistan: The Forgotten Crisis," *Refugee Survey Quarterly* 15, no. 2 (1996): 5; e ID., *The Fragmentation of Afghanistan: State Formation and Collapse in the International System* (New Haven: Yale University Press, 2002 (2ª)), pp. 48–52.

Although formally Afghanistan was never a colony, the Treaty of Gandamack imposed important limitations on its sovereignty. The Treaty of Gandamack was signed during the war (May 26, 1879) by the regent Yakub Khan. The Treaty of Gandamack was an authentic protectorate treaty which granted England international representation of Afghanistan. For their part, England paid a subsidy for the maintenance of the army. Further, Yakub Khan was forced to cede vast areas to the west of the Indus (Kurram, Pishin and Sibi), and accept that the Khyber and Michi passes would remain under British control. This meant the virtual secession of the Peshawar Valley and of other Afghan territories to the west of the Indus, which permitted the British, less than 15 years later, to set out the Durand line (RASANAYAGAM, Angelo, *Afghanistan: A Modern History* ..., cit., p. 8).

[18]RASANAYAGAM, Angelo, *Afghanistan: A Modern History* ..., cit., p. 10.

Although they were formally under British (and later Pakistani) sovereignty, these Pashtun tribal territories along the border with Afghanistan actually constituted a sort of buffer zone between British India (later Pakistan) and Afghanistan.[19] The British managed to govern these areas only indirectly, somewhere between the direct administration of the territories of the subcontinent and the Afghan protectorate at the outer margin of their zone of influence. In exchange for recognizing the colonial government and keeping the frontier passes open, the *maliks*, obtained not only a certain autonomy, but also allowances and subsidies to distribute among the tribes. While British domination lasted, there were numerous popular resistance movements in tribal areas which led in 1930 to the Pashtun nationalist movement.[20]

Pakistan inherited this special regime in the tribal areas which included its own legal system, which is in force even today. Pakistan also refused to disband the army in these regions after the *maliks* of Khyber, Kurram and the two Waziristans signed an agreement recognizing the new government, in exchange for maintaining the economic benefits they enjoyed. These agreements situated the bases of the current FATA (Federally Administered Tribal Agencies) located in the country's northwest on the border with Afghanistan. Each of the seven tribal agencies (Khyber, Bajaur, Mohmand, Orakzai, Kurram and the North and South Waziristans) has their own border guard militias and the state delegates administration of the law to tribal assemblies (*jirgas*) and to the *maliks*, or notables. In conclusion, the tribal zones were characterized by their non-submission to British domination first, as well as by their virtual autonomy from Pakistan.

The permeability of the Afghan-Pakistani border has lasted until today. Furthermore, the panorama has been complicated with the arrival of radical Islamism. As we will see further on, Pakistan has been supporting the Afghan Islamists since the 1970s. Over more than 30 years of conflicts in Afghanistan, the tribal areas in Pakistan have been used as a jumping-off point for war against the Soviets, and later as a base camp for the Taliban, who, currently, are endangering relations between Pakistan and Afghanistan, the stability of both and of the region in general, as well as the struggle against terrorism. There is no doubt that the support of President Musharraf was a key factor in the resurgence and success of the Taliban insurgency in Afghanistan, but it is no less certain that even if Pakistan had the will to throw out the Taliban from its frontier regions – as it seems to since the political changes which followed the 2008 elections and which ended the Musharraf era- they would have difficulty doing so, given that Pakistan's tribal areas in large part evade government control. Furthermore, the problem has begun to return in

[19]RUBIN, Barnett R., "Afghanistan: The Forgotten Crisis", cit., p. 5; e ID., *The Fragmentation of Afghanistan: . . .* , cit.

[20]ESPINOSA, Ángeles, "Las áreas tribales pakistaníes y la lucha contra el terrorismo," *Política exterior*, no. 116 (marzo–abril 2007): 52–56.

Pakistan, where a progressive radicalization of the frontier zones is occurring.[21] In this way, a problem as current as the ungovernability of the Afghan-Pakistani border has its roots in an anomaly from the colonial period.

Afghanistan's function as buffer state allowed it a relative stability between 1880 and 1975, at which point, as we will see, its status as buffer state began to crack, leading to a long process of disintegration, which has lasted until today.[22]

4.2 The Equilibrium and Disequilibrium of Afghanistan During the Cold War

Notwithstanding the undoubted weakness of the state, during the 40 years, from 1933 to 1973, coinciding with the reign of Zahir Shah (the last king of the Durrani dynasty), Afghanistan was one of the most peaceful and stable countries in Asia. In 1934 it joined the League of Nations. It remained neutral during the Second World War. During the post-war era, its position as a member of the international system was widely accepted. It avoided war with its neighbours and internally it was free of violence and mass killings. This was perhaps the reason why worrying signs, which as far back as the seventies foretold convulsions in Afghanistan, went unnoticed.[23]

During the reign of Zahir Shah, there were two phenomena which proved decisive for the future of Afghanistan: a progressive dependence on foreign aid along with the country's entry into the Soviet Union's sphere of influence, and a growing loss of legitimacy of the state.

Because of the lack of economic development, the fiscal bases of the State were extremely weak. This resulted in a need to resort to external help, not only of a financial kind, but also technical assistance to carry out reforms and investment in infrastructures.[24] The consequences were as follows: while in 1953 only 7% of state expenses were financed with foreign funds, by 1963 this accounted for 49% of the state's requirements.[25] When Daud came to power in 1973, Afghanistan had become a client-state ("*rentier*" o "*allocation*" state), dependent – for more than 40% of its budget – on income from outside ("revenue accruing directly from abroad").[26]

---

[21]On the situation in tribal areas from Pakistan's independence until today, ESPINOSA, Ángeles, "Las áreas tribales pakistaníes y la lucha contra el terrorismo", cit., pp. 51–62 *passim*.

[22]GIUSTOZZI, Antonio, "Afghanistan: The Patrimonial Trap and the Dream of Institution-Building", cit., pp. 71–2.

[23]MALEY, William, *The Afghanistan Wars*, cit., p. 6.

[24]GIUSTOZZI, Antonio, "Afghanistan: The Patrimonial Trap and the Dream of Institution-Building", cit., pp. 73–4.

[25]RUBIN, Barnett R., *The Fragmentation of Afghanistan: ...*, cit., p. 296.

[26]This income included both external aid and the sale of natural gas to the USSR, which began in 1968. (RUBIN, Barnett R., "Afghanistan: The Forgotten Crisis", cit., p. 7).

For a long time, Afghanistan tried to avoid involvement in the Cold War. Notwithstanding, from the mid-1950s, although the government received aid from both blocks, there was in increasing influence by the Soviet Union in the country. Thus, from 1956 to 1978 the sum of Soviet economic and military aid was nearly five times greater than the American.[27] It should be noted as well that a considerable part of the latter was granted in the 1950s, after which Washington lost interest. Thus, at the peak of the Cold War Afghanistan entered the sphere of Soviet influence.

The substitution of the internal sources of income for external financing tends to have extremely negative effects on the model of relations between society and the state, at the last, on the social legitimacy of the latter. When the state does not depend on the economic activities of its citizens in order to finance itself, the political leaders are not constrained by the need to satisfy the citizens' demands, to "negotiate" with them or to account for their policies. Thus they can ignore the promotion of economic development and democracy. "Unfortunately, the need to deliver justice can be short-circuited by foreign support, which injects wealth for the state's direct use". The consequence of this is a loss of the state's legitimacy. This is what happened in the final years of Afghanistan's monarchy, and during Daud's Republic.[28]

In the final period of Zahir Shah's reign (1964–1973), known as the "New Democracy", there was an attempt made to establish a constitutional monarchy. Nonetheless, the attempt failed because no authentic social or political reforms were implemented, nor were the institutions modernized. The bureaucracy remained completely dysfunctional. Its civil servants were among the worst paid in the world. As a result, corruption was extensive. In rural areas, the traditional structures of agrarian society prevailed against the government's attempts to impose its power. The weakness and de-legitimization of the state reached its maximum expression with the famine suffered by the country in 1972. As Sen explains, the famines were not so much a result of a lack of food, but of the state's inefficiency. Furthermore, in the Afghan case, the response by institutions and the indifference of those governing toward human suffering were truly deplorable. In these circumstances, it is not surprising that the weakened regime of Zahir Shah was ousted by a coup d'etat in 1973, headed by the former Prime Minister Daud, supported by the army and by the early Communist parties of the country's interior. Afghanistan was declared a republic and Daud governed as President.

During Daud's regime, the legitimacy of the state – as Maley continues – deteriorated even more due to the disappearance of the monarchic system, which at least offered "the semblance of a rule for peaceful succession to political

[27]Between 1956 and 1978, Afghanistan received some 533 million dollars in economic aid from the United States and about 2500 million from the Soviets in economic and military aid (RASHID, Ahmed, *Descent into Chaos* ..., cit., p. 8).

[28]MALEY, William, *The Afghanistan Wars*, cit., "State and society" and "The decline of state legitimacy: 1964-78", pp. 13–7. Quote from p. 15.

authority".[29] Daud tried to concentrate all power in his own hands. He tended toward bureaucratic centralization but like his predecessors, he was incapable of creating institutions.

Under Daud's regime, we once again see the "paradox of weak states": the use of repression, in particular against the incipient fundamentalist Islamic movement, and the attempt to install a powerful central government which only hides the weakness of the state, its institutions and its internal cohesion.

In April 1978, the "Saur Revolution" led by officials of the army inspired by Marxist ideology, put an end to the Daud government.

The new regime attempted to impose a revolution from above, introducing ample territorial reforms (redistribution of land and modification of the rents system), educational ones (formal, obligatory and secular education), and reforms related to the social status of women, leading to a strong backlash against them. The reforms clashed with the deepest structures of Afghan society. Thus, the agrarian reform ignored the needs of the nomads and the importance of access to water and seed. And the reforms related to women were seen as an attack against essential aspects of Islam. Furthermore, although there were well designed and carefully applied reforms, the attempt to introduce them in an accelerated manner through the party's bureaucratic apparatus, instead of explaining them and working through community structures, was what led to the discontent and the surge in a growing number of conflicts and armed insurrections. As Rubin explains: "the choice the peasants were given was not between domination and exploitation on the one hand and freedom and equality on the other. Their choice was between leaders whom they knew, with whom they shared much, and leaders whom they did not know, who believed in an alien ideology and who showed by their actions that could not be trusted".[30] Furthermore, the decision to use force in order to suffocate the revolts was counterproductive, as it reinforced opposition to the regime.

4.3 The Afghan – Soviet War: Afghanistan as a Battleground Between the Superpowers

The internal struggles among the varying PDPA factions, along with the failure of the economic, social[31] and political reforms, and the strength being gained by the mujahedin insurgency alarmed the Soviets, who invaded Afghanistan on Christmas Eve of 1979.

---

[29]MALEY, William, *The Afghanistan Wars*, cit., pp. 16–7.

[30]RUBIN, Barnett R., *The Fragmentation of Afghanistan: . . .*, cit., p. 119.

[31]For a detailed analysis of the social reforms undertaken (women's empancipation, literacy campaigns and agrarian reform), of the results obtained as well as the difficulties in implementing them, particularly in rural areas, throughout the entire period of the PDPA regime, see GIUSTOZZI, Antonio, *War, Politics and Society in Afghanistan 1978-1992* (London: Hurst&Company, 2000), pp. 20–32.

Afghanistan suffered the consequences of the superpower showdown during the Cold War particularly intensely, due to the Soviet invasion and the subsequent war. Rubin claims that the rupture of the distension in the 1970s among the superpowers was what led to the war breaking out. Until the mid 1970s, the United States and the USSR had competed to exercise their influence over the Afghan regime, but both had supported it, instead of supporting the political factions seeking to replace it. Nonetheless, following Daud's coup d'etat in 1973, the increase of tension among the superpowers affected their attitudes to the internal political forces in Afghanistan. "Both, the URSS and Pakistan, the latter with U.S. support, increased their aid to political groups challenging the Afghan regime, the Communists and Islamic movements,[32] respectively". The Afghan Soviet War had become part of the "intense endgame of the Cold War".[33]

Some of the causes of the war were likely internal: the increasing use of repression by some of the political leaders, and their search for foreign support for this, along with rivalries and struggles among the profoundly divided political elites. However, the scale and intensity which the conflict reached were the result of the intervention by foreign powers, specifically the USA and the USSR, though not only. As Alan Phillips points out, the principle cause of the war was neither the ethnic nor the religious conflicts. Prior to 1979 there were tensions among the ethnic groups, but not war. "In Afghanistan the conflicts were primarily political, not ethnic and religious, and were fuelled by superpowers politics".[34]

Certainly, Afghanistan was a weak state, but this circumstance alone would not explain the country's decomposition. Although the weakness of the state implies a proclivity to failure, the move from one state to another, as Rotberg points out, is not inexorable. Some weak states succumb to failure or collapse, while others, including in more difficult circumstances, continue being weak but without ever being destroyed. For Rotberg, the reason for this different evolution lies exclusively in internal factors, and more so in the structural and institutional deficiencies in which he calls "the Hand of Man": the destructive leaders, corrupt and/or repressive behaviours of the leaders and the political elites. Therefore, the failure of the state can be avoided if measures are taken concerning good governance, in order to detain the deterioration.[35] As I see it, the causes for the failure of the state lie not only, and occasionally not even principally, in internal factors. The systemic factors of an

[32]Pakistan began supporting the Afghan Islamists in 1975 in order to pressure Daud and as a response to the support he had given to the Pakistani Pashtun and the Marxist Baluchis. Among the Islamists were Gulbudin Hekmatyar, Burhanudin Rabani and Ahmad Shah Masud, who would later become important Mujahedin leaders in the war against the Soviets. During this period, a model of behaviour emerged which has persisted until today "with Pakistan using Islamic movements in Afghanistan to pressure the authorities in Kabul or as a means to try and install a more friendly government" (RUBIN, Barnett R., "Afghanistan: The Forgotten Crisis", cit., p. 11).

[33]RUBIN, Barnett R., "Afghanistan: The Forgotten Crisis", cit., pp. 6, 10 and 15. Quote from p. 10.

[34]PHILLIPS, Alan, "Preface" to JAWAD, Nassim, *Afghanistan. A Nation of Minorities*, cit., p. 5.

[35]ROTBERG, Robert I., "The Failure and Collapse of Nation-States: ...", cit., pp. 10 and notes.

international nature also play a decisive role[36] and the case of Afghanistan is fine proof of this. The Afghan – Soviet War established the bases for the breakdown of Afghanistan as a result of the superpower policies and the increasing implication of ambitious regional powers, principally Pakistan and Iran, in Afghanistan's internal affairs; an intrusion with fatal consequences for the country's future and the region in general. The invasion and subsequent war against the Soviets strengthened the Islamists and gave them, initially at least, legitimacy in the eyes of the people. The arms and the money from the United States and its European and Arab allies allowed them to create a power infrastructure based on violence, which was very difficult to take apart.

As Rashid underlines, for the Afghans the Soviet invasion represented "another attempt by outsiders to subdue them and replace their time-honoured religion and society with an alien ideology and social system". This led to the initial popular support for the mujahedin who, financed by the United States, became the "anti-Soviet shock troops".[37]

Islam played a fundamental role as a base for the resistance, granting it legitimacy.[38] The opposition against communism and the Soviet occupation was presented as a *jihad* (as opposed to the Arabic term *harb* or the Persian *jang*, which mean war). Even the words *mujahid* and the plural *mujahedin*, which became generic terms to designate Afghan resistance and its various components stem from the same Arabic root as *jihad* and mean "those who practice *jihad*".

Now, the great majority of those who initially took up arms against the Communist regime and the USSR were not defending an ideological Islam ("were not Muslim intellectuals"), but instead practiced "village Islam" which provided the moral basis for opposition to the atheism represented by the government and the Soviets. Initially, the war fought by these groups was tribal, pre-modern and traditional, as opposed to the contemporary ideological *jihad*.[39]

The foreign aid contributed significantly to the escalation of the war and to the radicalization of the resistance. The United States were the principle source of financing for the Afghan resistance. Saudi Arabia and other Arab countries also played an important role. According to some sources, the contributions form Saudi Arabia equalled those provided by the United States, including direct donations from private Arab sources to certain resistance parties, particularly that of Sayyaf,

---

[36]A broad discussion of the causes for the failure of states in FERNÁNDEZ RUIZ-GÁLVEZ, Encarnación, *¿Estados fallidos o Estados en crisis?* (Granada: Comares, 2009), chapter III, pp. 98–120.

[37]RASHID, Ahmed, *Taliban:* . . . , cit., p. 13. See also MALEY, William, *The Afghanistan Wars*, cit., "Popular mobilization in Afghanistan", pp. 58–66.

[38]MALEY, William, *The Afghanistan Wars*, cit., "Islam as a basis of resistance", pp. 58–60.

[39]MALEY, William, *The Afghanistan Wars*, cit., "Grassroots resistance and tribal warfare", pp. 60–2.

In the same line, Rashid points out that in Kandahar "the struggle against the Soviets was a tribal jihad led by clan chiefs and *ulema* (senior religious scholars) rather than an ideological jihad led by Islamicists" (RASHID, Ahmed, *Taliban:* . . . ,, cit., p. 18).

Ittihad-i Islami (Islamic Unity), a small group with a small social base, but which many commanders joined due to the large sums of money and arms which they had available.

Pakistan was the neighbouring state which played the key role in supporting the Afghan resistance (more specifically, as we shall see, specific resistance groups), helping them receive arms and providing training and logistic support.[40]

The United States headed up the opposition to the Soviet invasion and was the principle backer of the Afghan resistance, from the point of view of financing, but Pakistan was the protagonist of how the war was conducted. International aid reached the mujahedin through Pakistan. The Pakistani army were the principle route of transmission for it. In particular, the Pakistani secret service (ISI) directed the Afghan war. The United States gave too much freedom to the ISI to distribute the weapon of American funding, because they assumed that the ISI knew and understood Afghanistan and would have valuable contacts among the resistance groups. The result was that the United States was badly informed regarding the popularity of the deposed king Zahir Shah among the refugees, and about the true support enjoyed by some of the hard-line Islamic resistance groups.[41]

In its control of the Afghan war, Pakistan privileged some groups to the detriment of others. At the end of 1980 there were over 40 Afghan political parties (liberals, social democrats, nationalists and Islamic) registered with the Pakistani authorities. Nonetheless, General Zia's military regime permitted only seven exiled Afghan political parties to act from Peshawar and receive aid from the CIA. The seven were Sunni Muslim parties.[42] In principle, although not actually so in practice, the refugees were asked to join one of these parties in order to receive aid from the Pakistani program for refugees. Zia did not permit nationalist, democratic or left-wing secular parties to operate from Pakistan. The king Zahir Shah and his close family were not authorized to visit Pakistan. Furthermore, Zia insisted that the parties refer to the war as a *jihad* and not as a national liberation movement.[43] Further, the ISI favourites who received the lion's share of the aid were the more extremist groups, particularly Hekmatyar's *Hizb-e-Islami* (Party of Islam) (and this despite their lack of military results). Rabani's party, *Jamiat-i Islami* (Islamic Society) also received substantial assistance from Pakistan, although they sometimes withheld the aid.[44] The two most important regional commanders,

---

[40]MALEY, William, *The Afghanistan Wars*, cit., "The role of Pakistan", pp. 66–76.

[41]WEINBAUM, Marvin G., *Pakistan and Afghanistan: Resistance and Reconstruction* (Boulder: Westview Press, 1994), p. 32.

[42]A detailed explanation of same in RUBIN, Barnett R., "Afghanistan: The Forgotten Crisis", cit., pp. 17–9. Also in MALEY, William, *The Afghanistan Wars*, cit., pp. 63–4.

[43]RASHID, Ahmed, *Descent into Chaos* . . . , cit., p. 10.

[44]"The ISI – writes RASHID, *op. cit.*, p. 10 – used the CIA cash and arms as bribes to keep the Mujahedin's parties in line".

Masud and Ismail Khan, were associated with this party. In exchange, the two more moderate parties, led by Sufis, received little foreign aid.[45]

To summarize, Pakistan contributed to the fact that radical Islamism headed up the resistance to the Soviets. Prior to the war the Islamists had little support in Afghan society, but the CIA money and arms, along with the support of Pakistan allowed them to consolidate themselves and acquire enormous political power.[46]

This had a very prejudicial effect on the situation of women in Afghanistan and the West bears some responsibility for this for having supported the Mujahedeen, and for having permitted the Islamists to become the sole opposition to communism recognized by the West. As a result, the women's cause was sacrificed. And this occurred during the same period in which the repression of women by the Iranian revolution was being harshly criticized by the same Western circles who praised the Mujahedeen as heroes.[47]

In addition, the anti-soviet *jihad* mobilized many radical Muslims from very diverse countries. These were the so-called "Afghan Arabs", although they were not Afghans, but foreign combatants, and not all of them were Arabs. Recruitment of these volunteers was organized by the ISI and by Saudi Arabia's secret services, and from 1986 had the support of the CIA as well. These volunteers attempted to propagate Wahhabism and Pan-Islamist ideas in Afghanistan. The Wahhabism had little popular support in the country. Attempts by the radical Arabs to promote it caused quite a bit of conflict in the Afghan resistance movement and had little success. Nonetheless, they recruited a small number of Pashtun supporters, due to abundant funds and arms which they had available. Among the foreign mujahedin Osama Bin Laden played an important role. In fact, this was the base from which he would later create the terrorist organization Al Qaeda, gathering the Arab mujahedin who had remained in Afghanistan and Pakistan following the Afghan-Soviet War and recruiting thousands of new militants from all over the world.[48]

Furthermore, the anti-Soviet Jihad and later evolution contributed to the creation of a new military class of warlords and strong men, as well as to their growing

---

[45]This was the National Islamic Front of Afghanistan, led by Gailani, who was the most liberal and the closest to the old monarchy regime of the parties recognized by Pakistan; and the Afganistán National Liberation Front, lef by Sibgatullah Mujadedi. The latter was the party which Hamid Karzai joined. Sibgatullah Mujadedi was, in Rashid's words, *op. cit.*, p. 10, "[a] moderate man, he drew his support from tribal leaders who were Afghan nationalists rather than mullahs and militants, and became a fierce critic of the radical leaders who were favored by the ISI". Rubin describes him as very well known but with no real power.

[46]RASHID, Ahmed, *Taliban:* ..., cit., p. 19.

[47]MANN, Carol, *Femmes afghanes en guerre* (Éditions du Croquant, 2010), pp. 145–9.

[48]RASHID, Ahmed, *Taliban:* ..., cit., Part 2, chapter 10, "Global Jihad: the Arab-Afghans and Osama Bin Laden", pp. 128–140; y RASANAYAGAM, Angelo, *Afghanistan: A Modern History* ..., cit., chapter 16, "Holy War, Unholy Terror", pp. 213–249.

autonomy vis-a-vis political controls.[49] These militarized leaders replaced the traditional leaders of Afghan society, becoming the "new Khans".[50] These were the new "political" elites -in the less noble sense of the expression- promoted by war, who in the first stages competed for the *jihad* leadership and for control of resources (arms and money) which reached the combatants from abroad. Later, once the anti-Soviet war was over and Najibullah's Communist regime was toppled, the heyday of the warlords arrived[51] as they fought amongst themselves in an anarchic war the only motivation for which was a struggle for power – again in the less noble sense of the word.

Within the context of the refugee camps in Iran and Pakistan, a new generation of Afghans emerged who were unprepared to carry out the habitual economic activities (farmers, shepherds, businesspeople). This "disruption to human capital formation" simply fed the cycle of violence, given that "it is often easier to train unskilled youths to fight than to farm".[52] As we will see further on, the emergence of the Taliban was, in part, a result of this situation.

In summary, in Afghanistan the war caused a de-structuring at all levels of politics, of the economy, of society, as well of the culture.[53] However, as highlighted by analysts and certain official UN documents, the collapse of Afghanistan in the 1990s could have been avoided if the international community had been committed to the consolidation of peace in Afghanistan. This question is analyzed in the next section.

4.4 International Indifference and the Collapse of Afghanistan in the Post-Cold War

Unlike other conflicts linked to confrontations among the blocks, which could be resolved or at least a solution worked towards after the end of the Cold War, for example in Cambodia or Mozambique, this was not the case in Afghanistan.

---

[49]GIUSTOZZI, Antonio, *Empires of Mud. War and Warlords in Afghanistan* (London: Hurst&Company, 2009): Within the Afghan context, the warlords emerged from the Jihadist movement (pp. 43–51) and also as a consequence of the evolution of militias recruited in order to confront the insurgency (pp. 53–68). Among the foremost was Ismail Khan, in the Herat region (ch. 14–21), while the most important warlord among the second group was Dostum in the north of the country (ch. 6–13). Giustozzi considers Masud not to have been a warlord per se, but rather a hybrid between warlordism and the guerrilla war (ch. 22).

[50]GIUSTOZZI, Antonio, *Empires of Mud. War and Warlords in Afghanistan*, cit., pp. 33–4.

[51]GIUSTOZZI, Antonio, *Empires of Mud. War and Warlords in Afghanistan*, cit., pp. 69–85.

[52]MALEY, William, *The Afghanistan Wars*, cit., pp. 154–5.

[53]MALEY, William, *The Afghanistan Wars*, cit., p. 153; "Effects of the war on Afghanistan", pp. 154–9; y p. 167. See also, on cultural de-structuring JAWAD, Nassim, *Afghanistan. A Nation of Minorities*, cit., pp. 26–7.

Certainly, the country's panorama is particularly complex, but in addition, the international community have made repeated errors and omissions since the end of the Afghan-Soviet war until the present day which have contributed to the constant worsening of circumstances in the country, leading to the extremely serious situation which exists today.

Following the Soviet retreat in 1989, the United Nations and the international community in general were uninvolved in consolidating the peace in Afghanistan. This was for two main reasons. The first is the finalization of the bipolar conflict which had led the United States, upon winning the Cold War, to lose interest in Afghanistan and in the region in general.[54] In the 1990s, the United States practiced a policy of "Russia first", which led them to ignore Central Asia (apart from the dismantling of the Soviet nuclear warheads in Kazakhstan).[55] The politics of George Bush (the elder) regarding Afghanistan was erratic at best, always taking great care not to endanger Gorbachov's position. Further, Bush had left the Afghan question in the hands of Secretary of State James Baker and of the Ambassador in Islamabad, Oakley, known in the region as the "Viceroy". This delegation of political responsibility to the United States embassy in Pakistan would have serious consequences in coming years.[56] The second reason was that the international community had other priorities in the nineties: Iraq and the Balkan crisis. The international community did not take a serious interest in Afghanistan again until September 11th, when suddenly the Central Asian country became a strategic priority for the West.

The Geneva Accords of 14 April 1988, reached through the mediation of the Secretary General of the United Nations, achieved the complete withdrawal of the Soviets from Afghanistan on 15 February 1989. The Accords hoped to recuperate international cooperation regarding Afghanistan,[57] and foresaw not only the withdrawal of the Soviet troops, but also a cessation of aid to the mujahedin based in Pakistan. However, this aspect of the accords was never fulfilled. The United States and their partners continued to aid the mujahedin (Masud in particular received a great amount of aid from the United States in 1990–1991), just as the Soviet Union continued to support the Communist regime in Afghanistan. This flow of financing from abroad for the combatants did not help with the pacification of the country. The Geneva Accords ended the war against the Soviets, but the war continued between the mujahedin supported by the United States, Pakistan, Saudi Arabia and other Arab and Islamic countries and the Najibullah regime backed

[54]RASHID, Ahmed, *Descent into Chaos . . .*, cit., p. 11.

[55]KHANNA, Parag, *The Second World . . .*, cit., p. 69.

[56]MALEY, William, *The Afghanistan Wars*, cit., "US disengagement", pp. 178–80.

[57]RUBIN, Barnett R., "Afghanistan: The Forgotten Crisis", cit., p. 29; e ID. *The Search for Peace in Afghanistan: From Buffer State to Failed State* (New Haven: Yale University Press, 1995), Part Two, "Negotiating the Geneva Accords", pp. 31–91.

by the USSR. In Rubin's words, "[t]he Accords ignored the domestic conflict in Afghanistan and failed to reinstate superpower cooperation. They assured Soviet troop withdrawal on a fixed schedule while leaving in place a proxy war".[58]

Despite the continuing war in Afghanistan, Rubin sustains that during Najibullah's interregnum (1989–1992), coinciding with the period in between the end of the Cold War and the disintegration of the Soviet Union, there was a certain amount of international cooperation among the superpowers regarding Afghanistan.[59] This facilitated the elaboration of a United Nations Plan for Transition in Afghanistan (Plan by the Secretary General Perez de Cuellar in May of 1991). Further to this, on the 13 September 1991, the United States and the Soviet Union agreed to stop sending aid to the various groups.[60]

Nonetheless, following the collapse of the Najibullah regime in April 1992, the attempt to establish a coalition government with the participation of all groups failed, as they were unable to reach a power-sharing agreement among the various factions. At that moment, the process of replacing politics with violence caused by the Afghan-Soviet war – in which the two super-powers had provided sophisticated arms and massive amounts of aid to numerous groups – was already well advanced. The flow of arms empowered the new elites, but without training them to exercise political power. In 1992, there was no party or unified group capable of exercising legitimate power in the entire Afghan territory.[61] As Rubin stated "When state institutions unravel, and armed factions emerge as the main form of collective action, interim governments offer no quick solution to the problem of political order. No government can compensate for the dissolution of the state".[62]

Simultaneously, a war of anarchy was unleashed between opposing factions of older Mujahedin for control of the capital.

Rashid ties the civil war in the 1990s to ethnic rivalries among the Pashtuns and non-Pashtuns. The civil war was provoked, in large part, by the fact that Kabul did not fall into the hands of the Pashtun parties, but into those of the Tajiks and the Uzbeks. This was "a devastating psychological blow", because for the first time in three centuries, the Pashtuns had lost control of the capital.[63] Rubin points out that at first, when the non-Pashtun forces, particularly in the north, joined together to take Kabul and repel Hekmatyar's attacks, the conflict seemed like a Pashtun

---

[58]RUBIN, Barnett R., *The Search for Peace in Afghanistan: ...*, cit., p. 33.

[59]RUBIN, Barnett R., *The Search for Peace in Afghanistan: ...*, cit., chapter 7, "Cooperation Between the Superpowers", pp. 96–111.

[60]Joint declaration by the new Soviet Minister for Foreign Affairs and Secretary of State James Baker, Moscow, 13 September 1991.

[61]RUBIN, Barnett R., "Afghanistan: The Forgotten Crisis", cit., pp. 3, 6 and 15–22.

[62]RUBIN, Barnett R., "The Failure of an Internationally Sponsored Interim Government in Afghanistan," in *Between States: Interim Governments and Democratic Transitions,* eds. SHAIN, Yossi and LINZ, Juan J. (Cambridge: Cambridge University Press, 1995), p. 236.

[63]RASHID, Ahmed, *Taliban: ...*, cit., p. 21.

versus non-Pashtun conflict. Notwithstanding, as time passed, the ethnic structure of the conflict changed, making it clear that the struggle for power, more than ethnic relations, was the fundamental cause of the conflict.[64]

Further, the civil war of the 1990s was fed by the state's neighbours (Pakistan, Iran, India and Uzbekistan, and to a lesser degree Saudi Arabia, Russia and Turkey) who continued to support the groups in the conflict.[65]

From 1992 to 1996 Afghanistan became a collapsed state. This was the heydey of the warlords.[66] The rule of law had disappeared completely. There were neither national institutions nor the basis for political legitimacy. There were constant violations of human rights and of international humanitarian law, and what is more serious, there were no functioning institutions in the country for the protection of rights. There was no constitution, no legal system, no tribunals, no police and no army. The country was in the hands of a series of regional armed quasi-authorities, none of which satisfied international standards, nor even traditional Afghan standards. The insecurity was comprehensive.[67]

While this was occurring in Afghanistan, what was the attitude among the international community? Following the death of the Soviet Union, the country lost its strategic importance for the great powers. The United States, which had begun to move away from the Afghan problem following the Soviet withdrawal in 1989, was even more indifferent following the fall of Kabul to the mujahedin in 1992. The Afghan people were left on their own. The CIA left the management of its Afghan policy to its regional allies, Pakistan and Saudi Arabia, two states who had supported the more extremist mujahedin and who had sabotaged the possible resurgence of a national Afghan government. The strategic absence of the United States permitted countries in the region, including the newly independent republics of Central Asia, to support the rival warlords, which contributed to the intensification and prolonging of the civil war.[68] For nearly 2 years, there was no Afghan peace initiative from the United Nations, and when they finally approved a new goodwill mission,[69] the superpowers provided none of the financial resources. The UN political mission had a budget of only one million dollars. "Countries that paid thousands of millions for the war today can hardly find a million for reconstruction".[70]

Underlining the importance of the schemes for peace consolidation, in order to prevent future conflict, the Report of the High Level Panel on Threats, Challenges

---

[64]RUBIN, Barnett R., "Afghanistan: The Forgotten Crisis", cit., pp. 22–4.

[65]RUBIN, Barnett R., "Afghanistan: The Forgotten Crisis", cit., p. 31.

[66]RASHID, Ahmed, *Taliban: . . .*, cit, pp. 21–2; e ID., *Descent into Chaos . . .*, cit., pp. 12–3.

[67]RUBIN, Barnett R., "Afghanistan: The Forgotten Crisis", cit., pp. 3 and 6; and Amnesty International, *Afghanistan: A Human Rights Disaster*, 1995.

[68]RASHID, Ahmed, *Taliban: . . .*, cit, p. 175–176; ID., *Descent into chaos. . . .*, cit., p. 11.

[69]The Mestiri mission was authorized by the United Nations General Assembly, "Emergency International Assistance for Peace, Normalcy and Reconstruction of War-Stricken Afghanistan", A/RES/48/208, December 21, 1993.

[70]RUBIN, Barnett R., "Afghanistan: The Forgotten Crisis", cit., pp. 31–5. Quote from p. 35.

and Change pointed out that if the Security Council had truly made an effort to consolidate the peace in Afghanistan at the start of the decade of the 1990s, many lives could have been saved, perhaps the Taliban would never have reached power and Al Qaeda could never have found a safe haven in that country.[71]

4.5 Rise and Rule of the Taliban: From Anarchy to Oppression

The Taliban emerged in the south of Afghanistan where there is a Pashtun majority, at the end of 1994, as an alternative to the fragmented politics, the disorder and the anarchic violence which reigned all over the country. The older mujahedin leaders were divided into factions who fought amongst themselves and were dedicated to criminal activity. In these circumstances, the Taliban emerged as an Islamic reform movement. The Taliban believed that they were launching a *jihad* by struggling against the cruel and corrupt warlords, defending the population from the abuses committed against them. This explains the initial popularity the Taliban enjoyed among the Afghan population, in contrast to their rejection of the warlords. The Taliban brought relative peace and security to the zones under their control. The belligerent tribal groups were crushed, their leaders executed, the population disarmed and the highways unblocked.[72]

Notwithstanding, the pacification of the country by the Taliban was limited. On the one hand, the civil war between the Taliban and the anti-Taliban forces of the United Front (FU) continued throughout the period they were in power. On the other hand, the Taliban regime followed an increasingly repressive and totalitarian model, which is incompatible with true security in its fullest and most comprehensive sense of lack of fear, also vis-à-vis the repression practiced by the state.[73]

As Rashid underlines, the Taliban did not belong to any of the Islamic trends which emerged during the anti-Soviet war: radical Islamism along the lines of the Muslim Brotherhood of Hekmatyar and Masud, Sufism or traditionalism. It was specifically the breakdown and collapse of the legitimacy of these three streams involved in the vicious struggle for power which created an ideological vacuum which the Taliban occupied. Their principle religious and ideological inspiration was an extreme form of Deobandism[74] propagated in the Afghan refugee

---

[71] *A more secure world: Our shared responsibility*. Report of the Secretary-General's High-level Panel on Threats, Challenges and Change (A/59/565, December 2, 2004), par. 86.

[72] RASHID, Ahmed, *Taliban:* . . . , cit, pp. 1, 5, 22–5 y 87; ID., *Descent into Chaos* . . . , cit., p. 13.

[73] DÍAZ, Elías, *Sociología y Filosofía del Derecho* (Madrid: Taurus, 1980 (2ª); 1993 repr.), pp. 44–8.

[74] The Deobandists emerged in British India as a progressive movement who attempted to reform and unite Muslim society within the colonial context. The movement, which placed great

camps of Pakistan by the Pakistani Islamic parties, in particular Fazlur Rahman's *Jamiat–ul-Ulema* (JUI) (who were opposed to the *Jamaat-e-Islami*, the other large Islamic party of Pakistan).

During the 1980s the JUI established hundreds of madrasas along the Pashtun belt, all along the frontier with Afghanistan. There they offered young Pakistanis and Afghan refugees free education, food, shelter and military instruction. Given the great deficiencies of the Pakistani public education system, these madrasas represented (and still do) the only opportunity for many families to have their children receive something approaching an education. These madrasas were important for forming the Taliban leaders and a feeding source for combatants for the movement.

It is true that the orientation of these madrasas is far from the original reformist program of the deobandi schools, due to the Saudi Arabian influence which contributed to their financing as well as the influence of the *pashtunwali* (the tribal code of the Pashtun), or rather, a twisted version of itself, in its interpretation of *sharia*. Certainly, the deobandis tended to restrict the role of women, and opposed any form of hierarchy within the Muslim community, particularly tribal and feudal structures, and rejected Shiism, "but the Taliban were to take these beliefs to an extreme which the original Deobandis would never have recognized".[75]

It follows from this that the Taliban were in great measure the result of the refugee camps, of a feeling of rootlessness, of the militarized madrasas, and the lack of opportunities in the frontier regions of Afghanistan and Pakistan.[76]

The rise of the Taliban was tied as well to the interests of regional states (especially Pakistan and Turkmenistan) and to those of the Western oil companies (in particular Unocal, who were openly supported by the United States).

Disappointed by Hekmatyar, whom they had supported for two decades, Pakistan looked for a new Pashtun group capable of representing its interests in Afghanistan and of giving Pakistani commerce access to Central Asia; they found the Taliban. The Pakistani army's intelligence services became the principle supporters of the Taliban regime. The regime was recognized in 1997 only by Saudi Arabia, the United Arab Emirates and Pakistan. The latter convinced the two former to recognize the Taliban, and to support and finance them (money which they received through the ISI).[77]

On the other hand, following the breakdown of the Soviet Union, a new "Great Game" had begun,[78] this time for control of the energy resources of Central Asia,

---

importance on education, established madrasas all over India, the main one in Deoband. The Deobandi madrasas spread rapidly with the creation of Pakistan in 1947.

[75]RASHID, Ahmed, *Taliban:...*, cit, Part 2, chapter 6, "Challenging Islam: the new-style fundamentalism of the taliban", pp. 82–94. Quote from page 88.

[76]RASHID, Ahmed, *Taliban:...*, cit, pp. 1 and 23.

[77]RASHID, Ahmed, *Taliban:...*, cit, Part 1, chapter 1 "Kandahar 1994: The Origins of the Taliban" *in fine*, pp. 26–30; and RASHID, Ahmed, *Descent into Chaos...*, cit., pp. 14.

[78]This theory was launched by Rashid in an article published in 1997 (RASHID, Ahmed, "The New Great Game. The Battle for Central Asia's Oil", *Far Eastern Economic Review*, 1997) and later developed in his work *Taliban:...*, cit., Part 3, "The New Great Game", pp. 143–216. The

and the Caspian Sea (The Caspian region, made up of Kazakhstan, Turkmenistan, Azerbaijan and Uzbekistan). What emerged from the new "Great Game" was how and when to transport the oil and gas from the Central Asian-Caspian region to nearby markets, to then distribute in Asia and Europe. Basically, it was a matter of competition for oil and gas pipelines.[79] In this context, Afghanistan's instability created (and continues to create) a serious problem for the security of the potential oil and gas lines. In particular, the states and the oil companies needed to consider whether the Taliban would support the laying of the lines, as well as their ability to unify and stabilize the country, and based on this, decide whether to confront them or to try to attract them. Again, as had occurred in the nineteenth century, Afghanistan found itself at the centre of the "Great Game".

Rashid claims that the triumph of the Taliban was due in large part to the silence of the United States. In the post-Cold War era, the United States had not had a true political strategy in the region formed by Afghanistan, Pakistan, Iran and Central Asia. With regard to Afghanistan, he suggests, it was not that they had a secret plan; they simply lacked a plan of action. Taking this into account, in the American policy toward the Taliban, one can see various phases, conditioned by American interior politics, and by the attempt to deal with problems as they arose, and not by political strategy.

Between 1994 and 1996 the United States had politically supported the Taliban in an indirect manner, through their Pakistani and Saudi allies, as it was accepted that these would provide arms and financing for the Taliban. Above all, this was because Washington saw them as an anti-Iranian and anti-Shiite force who could counter the Iranian influence in Afghanistan and Central Asia.[80] During this stage, the United States deliberately ignored the fundamentalist Islamic program of the Taliban and their repression of women. Between 1995 and 1997 support for the Unocal project to construct a gas pipeline and an oil pipeline (*Central Asian Pipeline Project*) which would cross Afghanistan had raised a great deal of interest in the Taliban within Washington.

Instead of implementing real efforts to promote peace and lasting stability in Afghanistan, through understanding between the various groups and factions in order to form a broad-based government, the United States bet on the temporary stability which was implied in the Taliban conquest of the country. Once again, the inevitable task of consolidating peace in Afghanistan was delayed, and a shortcut was sought; the forced stabilization offered by the Taliban.

---

analysis which follows is based on those in this last work, except for express references to the contributions of other authors.

[79]AMIRIAN, Nazanín and ZEIN, Martha, *Irak, Afganistán e Irán. 40 respuestas al conflicto de Oriente Próximo* (Madrid: Lengua de trapo, 2007), pp. 107–111.

[80]On the error of this policy, Rubin wrote "Whatever the merits of the isolation policy towards Iran in the fight against terrorism, they incapacitate the US in Afghanistan" (RUBIN, Barnett, "US Policy in Afghanistan", *Muslim Politics Report* (New York: Council of Foreign Relations, 1997)).

North American policy regarding the Taliban changed in 1997, during Clinton's second term, following the fall of Kabul to the Taliban and the arrival of Bin Laden to the country (in 1996), and following the defeat of the Taliban at Mazar-i-Sharif, the fortress of the opposition in the country's north, in May of 1997. The latter created a growing scepticism in Washington regarding the ability of the Taliban to unify the country. As a result, the United States began to think of other ways to distribute the Turkmen gas. Furthermore, the efficient campaign by North American feminists against the Taliban also weighed into the policy change by the Americans.

Now it is true -continues Rashid- that not even in this second stage did the United States demonstrate a clear determination to put the brakes on the Taliban and on Bin Laden's activities in Afghanistan, and they were even less determined to confront the support given to the Taliban by the Pakistani army and by Saudi Arabia. At that stage, the CIA already saw Bin Laden as a threat but they nonetheless allowed him to continue using the entire country as a base of operations to mount a new, global terrorism structure, recruiting thousands of militant Muslims from various countries, and extending his operations worldwide.[81]

The Taliban's initial success in re-establishing order gave them a certain credibility, not only internally, but also internationally. Thus, the political mission of the United Nations in Afghanistan considered, at first, that the Taliban were a cohesive political force and that they would be able to reconstruct the institutions.[82]

Nonetheless, following the fall of Kabul, the Taliban showed growing centralizing and expansionist tendencies, but without necessarily strengthening the institutions. All power was concentrated in the hands of Mullah Omar, and Kandahar became the de facto capital of the country. Areas of the state system which did not interest the Taliban, such as the secular police, and the educational system, were abandoned and consequently, decayed even further. On the other hand, although increasingly dependant on foreign combatants, the army and the intelligence services were maintained, as well as some commercial relations. Progressively, from an internal point of view, the regime became a totalitarian system characterized by a religious darkness, the persecution of non-Pashtun ethnic groups, and the oppression of women. Further, its ties to the international illegal economy, and to broad-based terrorist groups, among them Al Qaeda, turned it into a destabilizing regime from an international point of view.

To summarize, the Taliban regime contained many of the characteristics of the so-called "rogue states", and should have been considered thus by the United Nations political mission to Afghanistan in 2000, several months prior to the military intervention in November 2001.[83]

---

[81] RASHID, Ahmed, *Descent into Chaos...*, cit., p. XLV and pp. 15–21.

[82] Along these lines, UN Special Mission to Afghanistan, Civil Affairs Unit, *Report on Administrative and Judicial Structures of Afghanistan: A First Inquiry*, Islamabad, 2000.

[83] In this sense, DUFFIELD, Mark, *Development, Security and Unending War. Governing the World of Peoples* (Cambridge, UK: Polity Press, 2007), pp. 154–5.

On the other hand, as Duffield states, the United Nations Strategic Framework for Afghanistan set out from the consideration that they were dealing with a "failed state". Thus they split from the former description of the country as an "impoverished war-torn society", characterized by the fragmentation of the community, the lack of social capital, the collapse of basic services, the disappearance of traditional mechanisms of protection, the absence of an effective government and the collapse of practically all state institutions, constituting an "emergency of governance". Under these circumstances, the only thing functioning was the criminal economy: the economic benefits of the war, poppy cultivation, drug trafficking and cross-border contraband.[84]

This concept of Afghanistan as a failed state justified the practical attribution of a predominant role in the area for development, regarding the political mission, and the assignation of a political function to aid, using it as a tool to construct "peace from below", at the community level, strengthening social cohesion, reconstructing civil society and contributing to behavioural transformation. The hypothesis was that where the state had failed, aid could succeed. The failure of the Afghan state justified the United Nations mission assuming the role of "surrogate government". Taking into account the later overthrow of the Taliban regime in 2001, Duffield points out that "in retrospect, this was an international government in waiting".

Similarly, from a specific moment, there was an attempt to use the aid to modify the Taliban's political conduct through the so-called principled engagement, although this supposed, in part at least, a contradiction of the view of Afghanistan as a failed state. It was the old carrot and stick system. Although the vital humanitarian aid was not conditional, other areas of assistance depended on the fulfilment of certain minimal standards, such as non-discrimination on the basis of gender among United Nations workers and access to all sectors of the population, in particular women, minorities and other vulnerable groups in need of aid. In this policy there was a tendency to distinguish between the "good" and "bad" parts of the Taliban regime. Thus, no direct assistance was given to the authorities in those areas where the principles of the Strategic Framework could be vulnerable, but the United Nations agencies continued to work in those sectors of public administration that provided essential services to the population in a non-discriminatory manner. The risk of this policy consisted in the fact that, more than a compromise of principles, it became an accommodating position regarding the Taliban, in particular in matters related to women, an area in which the Strategic Framework "conspicuously failed to develop a coherent or principled strategy of engagement".[85]

It is notable that the Taliban regime was considered "a rogue state" in one case, and a "failed state" in another, within the same United Nations context. It could be that often, the Taliban regime presented characteristics of both. Notwithstanding,

---

[84]UN, *Strategic Framework for Afghanistan: Towards a Principled Approach to Peace and Reconstruction*, New York, 1998.

[85]DUFFIELD, Mark, *Development, Security and Unending War*..., cit., pp. 142–150. Citations from pp. 146 and 150, respectively.

I believe it was above all a "rogue state", especially from the internal point of the intense repression suffered by their own population, although following September 11th the consideration became almost commonplace that, under the Taliban Afghanistan had been a failed state which had given shelter to Al Qaeda. With this as a basis, the entire debate about weak states has been distorted. The Afghan case went from particularity to paradigm. State fragility and terrorism were considered one and the same. In this manner a whole doctrine developed concerning failed states as a threat to global stability.[86]

4.6 The 2001 Invasion and Later Evolution

For over a decade, in Afghanistan the international community had repeatedly failed in prevention. In the long term, as suggested in the Report by the High Level Panel, the consolidation of peace in Afghanistan is what would have prevented September 11th. Nonetheless, progressively from 1989, the international community headed by the United States, misunderstood the drawn out civil war in Afghanistan. In the medium term, the United States' passivity had permitted the ascent and consolidation of the Taliban regime and the conversion of Afghanistan into a centre of operations for various terrorist groups and in particular Al Qaeda. Further, as Rashid highlights, in the short period of the months immediately before September 11th, the unsustainability of Afghanistan's internal situation, the intensification of terrorist activities in the country and the many indications that an attack was imminent should have alerted the Western powers.[87] Despite the intense surveillance efforts in the 2 years before 11 September 2001, "[t]he reach and sophistication of the al-Qaeda network were underestimated".[88]

Following the repeated omission of the "responsibility to prevent",[89] suddenly, on 11 September 2001 the international community was reminded of Afghanistan, but only in consideration of the fact that the Taliban regime constituted a threat to international security due to their support of Bin Laden. The invasion of

[86]A detailed analysis of this theory and of the objections which could be made about it in FERNÁNDEZ RUIZ-GÁLVEZ, Encarnación, *¿Estados fallidos o Estados en crisis?*, cit., "Estados frágiles: ¿amenaza para la paz y seguridad internacionales?", pp. 83–9.

[87]RASHID, Ahmed, *Taliban*..., cit., "New Foreword Post-Terrorist Attacks, September 2001", pp. ix–xv.

[88]RASANAYAGAM, Angelo, *Afghanistan: A Modern History*..., cit., p. 242.

[89]This terminology is used in the Report on *The Responsibility to Protect*. Nonetheless, the military intervention in 2001 in Afghanistan had nothing to do with this sort of intervention which the Report on The Responsibility to Protect deal with. The invasion of Afghanistan was not one with "the purpose of human protection"; instead, as we will see, its purposes were of another sort. Although the "responsibility to prevent", like the "responsibility to reconstruct" may be applied as analogy to these interventions which are different to those in the Report on *The Responsibility to Protect*, and with the same character of conditions of legitimacy of any hypothetical intervention.

Afghanistan, led by the United States, got underway in November of 2001. In order to understand the subsequent evolution, it is important to take into account that the stated aim of the so-called Operation Enduring Freedom was not to reconstruct Afghanistan and consolidate peace in the country, but merely to bring down the Taliban regime, eliminate the Al Qaeda terrorist network and arrest Osama Bin Laden. Operation Enduring Freedom began without the express authorization of the Security Council. Nevertheless, no sooner had the operation begun than the Security Council expressed its support through Resolution 1378 on 14 November 2001.[90] To carry out this operation, legitimate defence was invoked, although in this case this was not in keeping with the classic interpretation traditionally used in the field of moral, political and legal philosophy, nor in jurisprudence or the doctrine consolidated with regard to interpretation of art. 51 of the Charter. Using a terrorist attack to authorize the exercising of legitimate defence,[91] would imply the expansion and modification of the scope of the same and as it is established in art. 51 of the Charter. This modification would raise "serious and grave problems". In the classic interpretation, legitimate defence is exercised against the author state of the aggression, and not against the territory of the state which provided assistance, support or succour for a terrorist organization. This last constitutes, in principle, a violation of state sovereignty. On the other hand, the requisite for the immediacy of legitimate defence demands that the response come "while the aggressive act is still developing and once it has begun". If the state which is the victim of the terrorist attack reacts with armed force without immediacy, this would not be a case of legitimate defence, but rather a matter of "armed reprisals", which are prohibited by international law.[92] It then follows that the moral legitimacy and the international legality of the invasion were highly arguable.

Notwithstanding this, Rashid initially defended the legitimacy from a political point of view, given that "only external intervention could save the Afghan people from the Taliban and Al Qaeda".[93] Now, he claims that the invasion and fall of the Taliban regime should have been accompanied by a firm, serious and constant commitment by the international community toward Afghanistan and the Central and South Asian regions in general, in order to pull them out of their instability, poverty and under-development. This would have involved truly making an effort in the work of reconstructing Afghanistan, but also in contributing to national

---

[90]About these events, CARDONA LLORENS, J., "La Resolución 1386 (2001) del Consejo de Seguridad autorizando la fuerza internacional de asistencia para la seguridad de Afganistán: ¿Un paso más en el debilitamiento de las Naciones Unidas?", *Revista española de Derecho Internacional* LIII, no. 1 and 2 (2001): 227–8.

[91]This is one way of understanding two Security Council resolutions written immediately after September 11th (1368, from 12 September 2001, and 1373, from 22 September 2001), though they are ambiguous and contradictory.

[92]SANJOSÉ GIL, A., "Algunas reflexiones sobre el Informe del Grupo de Alto Nivel creado por el Secretario General y el futuro del sistema de seguridad colectiva de las Naciones Unidas", *Anuario de la Asociación para las Naciones Unidas en España*, no. 7 (2005): 313–317.

[93]RASHID, Ahmed, *Descent into Chaos* . . . , cit., p. XLIII.

reconstruction and democratic reform in Pakistan and all Central Asia, given that the problems in the region are so deeply interrelated that they need to be addressed as a whole. On the other hand, none of this actually took place. The United States ignored the consolidation of Central and South Asia. With regard to Afghanistan, during the first years following the invasion, when the large majority of the Afghan population received the foreign soldiers and their helpers with open arms, the international community had a great opportunity to help the Afghan people, but they did not take advantage of the situation. Once again, the international community and especially the United States had let the opportunity pass to consolidate peace in Afghanistan. The invasion of Iraq played a large role in this, given that an enormous quantity of resources were funnelled there which could have been used for the national reconstruction of Afghanistan. The war in Iraq was an enormous historical error, with effects which are still being seen in Afghanistan, given that the Taliban had re-emerged in strength and the country is on the verge of state collapse (for the second time in less than 20 years). Furthermore, Pakistan is suffering the consequences of its self-destructive policy and the double-game, in which the Pakistani intelligence services supported the Afghan Islamic extremists since the seventies, while the extremism and violence were becoming consolidated and propagating themselves on their own territory.[94]

From the above, it emerges that the disaster which we are witness to today in Afghanistan was not inevitable in 2001 when the EU invaded the country, but rather quite the opposite. This would explain the broad support by the greater part of the international community who accepted the intervention (despite questionable international-legal legitimacy) and even the hopes which, initially at least, were raised among the Afghan people. The causes for the failure may be found in the lack of EU interest, consistent with the initial aims of the invasion, in the political and economic reconstruction of the country, along with the de facto predominance throughout all these years of the United States over NATO and the United Nations.

With regard to this, one must take into account that the international involvement in Afghanistan since 2001 came as a result of two principle causes. On the one hand, Operation Enduring Freedom, which I referred to earlier, which was initiated without the express authorization of the Security Council by a coalition of States headed up by the United States. On the other hand, shortly after, the UN Security Council, based on Resolution 1386 of December 20 2001, authorized the establishment of the Afghanistan International Security Assistance Force (ISAF), heeding the petition by the signatories of the Bonn Accord, authorizing the deployment of a force under a UN mandate which would help to maintain security in Kabul and the surrounding area and which could gradually extend to other urban centers and areas. The ISAF was installed with the consent of the interim Afghan government and with authorization, backed by Chapter VII of the United Nations Charter, to use

[94] RASHID, Ahmed, *Descent into Chaos . . .*, cit., *passim.*

force in the fulfilling of its mandate.[95] Control and command of the ISAF was held initially by the United Kingdom and from August 11 2003 by NATO. For its part, Resolution 1510 of the UN Security Council, of the 13 October 2003, authorized the expansion of ISAF operations throughout the country.

The aforementioned Resolution 1386 also established a special UN political mission in Afghanistan.[96] Successive Security Council Resolutions have fixed the mandate for both: for UNAMA and the ISAF. Since the ISAF began to carry out combat operations (from 2003), this was considered by both the Taliban and the Afghan population as a contradiction, given that the job of the UN is to bring peace yet, notwithstanding this, they operate under the same mandate as the ISAF which carries out war activities.[97]

The lack of a basic consensus regarding the ultimate goal of the NATO mission (ISAF), is notable, which has generated heterogeneous strategies. Since 2001, even before NATO took control of ISAF, they had moved "from a strategy which pursued eliminating Al Qaeda [. . .], to objectives of democratization and post-conflict reconstruction, only to end up with a war of counterinsurgency, which in turn demands a broader plan of economic development and state consolidation".[98]

In any case, despite the modifications reiterated in the declared, pursued objectives, and although the international community spends its time speaking of reconstruction, development, electoral processes, etc. the fact is that Afghanistan has not been truly reconstructed either politically or economically. And this is entirely consistent with the United States' lack of interest in such an end.

The current Afghan government is weak and corrupt and does not have the majority support of the people. The warlords are incorporated into the government as well as into the Afghan security forces. The various electoral processes which have taken place in the country are full of irregularities and accusations of fraud and have contributed toward delegitimization the Karzai regime.[99] It is worth asking how the situation has ended up in this way, which was certainly not inevitable in 2001.

As Giustozzi highlights, following the expulsion of the Taliban from power, the country faced two interrelated challenges.[100] Given that the Taliban State had been

---

[95]CARDONA LLORENS, Jorge, "La Resolución 1386 (2001) del Consejo de Seguridad autorizando la Fuerza Internacional de Asistencia para la Seguridad de Afganistán: . . . ", cit., pp. 227–45.

[96]Initially called the United Nations Special Mission to Afghanistan (UNSMA), it was later known as the United Nations Assistance Mission in Afghanistan (UNAMA).

[97]MASADYKOV, Talatbek, GIUSTOZZI, Antonio and PAGE, James Michael, "Negotiating with the Taliban: Toward a solution for the Afghan conflict", Crisis States Research Centre working papers series 2, 66 (London: Crisis States Research Centre/London School of Economics and Political Science, 2010), pp. 1–6.

[98]ORTÍZ, Antonio, "La OTAN tras la cumbre Estrasburgo-Kehl," *Política exterior*, no. 129 (mayo–junio 2009): 62–66.

[99]GIUSTOZZI, Antonio, "Afghanistan: getting worse before getting better?", Writenet, Geneva, Switzerland, 2009, pp. 4–7 y 15–9. Available at www.unhcr.org/refworld/pdfid/4a9797572.pdf.

[100]GIUSTOZZI, Antonio, "Afghanistan: Political Parties or militia Fronts?", en ZEEUW, Jeroen de (ed.), *From Soldiers to Politicians. Transforming Rebel Movements After Civil War*, Lynne Rienner, Boulder/London, 2008, pp. 179–204 passim.

rejected by the groups and factions newly arrived in power, the State needed to be re-built from scratch. At the same time, it was necessary to transform the violent policies of the warlords into more peaceful electoral policies, and this within the context of an extremely impoverished society which survived thanks to the drug economy and international aid.

One must take into account that in 2001 those who took power were armed, anti-Taliban opposition groups, principally *Jamiat-i Islami*, *Hizb-i Wahdat-i Islami* and *Junbesh,* who had mobilized all their resources in the fight against the Taliban, and who at the end of 2001 controlled thousands of armed men. The Bonn Accord of December 5 2001 was not a true peace accord, as it included only the various anti-Taliban factions, excluding the Taliban who were the other party to the conflict. This exclusion of the Taliban from the political process, as well as from the disarmament, demobilization and reintegration process (DDR), was one of the main errors committed by the West in Afghanistan after 2001 and possibly contributed to the later development of the insurgency from 2003.

The Bonn Accord established an interim government (Afghan Interim Authority) presided over by Hamid Karzai. The duration foreseen for this interim government was 6 months until the Loya Jirga could be constituted. It was established in June of 2002. Karzai was confirmed as president for the period of transition which was to last until the presidential and parliamentary elections that were held in October 2004 and September 2005 respectively.

In the initial period of transition (2001–2003), the governments presided over by Karzai were formed by coalitions of diverse groups. These were based on moderate monarchists (amongst whom was Karzai himself), and the Islamist militias of the *Jamiat*. The strongest faction within the Jamiat were the *Shura-i Nezar*, which included lieutenants of the fallen commander Masud, and in whose hands were the two most important ministries; Defense and the Interior. *Junbesh*, *Wahdat*, and *Ittihad* (Sayyad's party) also had an influence and a presence in institutions. "In practice, President Karzai governed the country through indirect rule, relying on multiple alliances with political factions, individual strong men, and tribal groups".[101] This impeded the strengthening of State authority and the construction of institutions. The incorporation of armed non-State group leaders into the State Administration in 2001 had corrosive effects on the weak Afghan State structure. Karzai was incapable of establishing a strong political leadership which would allow him to balance institutionalization and patronage; a political leadership which would have trouble emerging within the context of external intervention.[102] On the other hand, with the division of power not being institutionalised, this would have led to rivalry among the factions and constituted an incentive to maintain the armed militias. The Afghan factions were not sure of the future prospectives for their integration into the political system. As a result, they had no incentive to sacrifice the

[101]GIUSTOZZI, Antonio, "Afghanistan: Political Parties or militia Fronts?", cit., p. 186.

[102]GIUSTOZZI, Antonio, *Empires of Mud. War and Warlords in Afghanistan*, cit., p. 304.

privilege they derived from being able to use violence, in exchange for the potential benefits from constructing institutions. In fact, when after 2003 they tried to expel the military leaders from the Government, instability increased.

Furthermore, the armed, non-State groups wield enormous economic power, given that they are mafias who use force to protect and expand their interests and who operate both in the legal and the illegal economy. Their control of key institutions such as the police and the local authorities allows them to merge both sectors of the economy. One of their principal activities is drug trafficking. Not mainly the cultivation of poppies, but rather the provision of "security" for narco-traffickers, drug processing and trafficking across borders. Many warlords and strong men have become "politicians" (at least formally), gaining seats in parliament, but without ever abandoning violence or their illegal enterprises.[103]

Ultimately, one of the main causes of the failure of political reconstruction in Afghanistan can be found in the insufficient international pressure, as well as on the part of the Afghan government for having imposed demilitarisation on the armed forces. This may be explained by the composition of the Afghan governments which we discuss above. Regarding the international aspect, a fundamental problem in Afghanistan is that the coalitions formed during the invasion directed by the United States reinforced the power of the elites who did not share the objectives of consolidating democracy and reconstructing a legal economy.[104] One of the principle errors of the United States in Afghanistan was that they revitalized the warlords and drug barons, who, grouped under the "Northern Alliance", collaborated with the American troops in order to expel the Taliban. Furthermore, they had been joined by other leaders recruited by the CIA and converted into warlords, such as Hazrat Ali, whose men did not manage to capture Bin Laden in Tora Bora.[105]

Various authors claim that the United States had no intention of creating a strong government. Rashid seems to attribute this to circumstantial reasons. When the war ended (one should say "officially") in summer 2002, it was clear that the Karzai regime lacked resources to limit the power acquired by the warlords. In these circumstances, the implicit North American strategy was to leave Karzai in the capital with insufficient power, protected by the foreign forces, dependant on the warlords of the rural areas.[106] Herold on the other hand attributes it to deeper geostrategic reasons and states that: "Afghanistan represents merely a space that is to be kept empty [...] the space represented by Afghanistan in a volatile region of geo-political import, is to be kept vacant from all hostile forces. The country is situated at the centre of a resurgent Islamic world, close to a rising China (and India)

[103]GIUSTOZZI, Antonio, "War and Peace Economies of Afghanistan's Strongmen," *International Peacekeeping* 14, no.1 (2007): 75–89.

[104]BARANYI, S., *¿Estabilización o paz sostenible? ¿Qué clase de paz es posible después del 11-S?* (Madrid: CIP-FUHEM, 2006), pp. 18–9.

[105]RASHID, Ahmed, *Descent into Chaos*..., cit., chapter seven, "The One-Billion-Dollar Warlords", pp. 125–44.

[106]RASHID, Ahmed, *Descent into Chaos*..., cit., p. 133.

and the restive ex-Soviet Asian republics, and adjacent to oil-rich states". Therefore, no real effort has been made to reconstruct the country politically or economically. The intention has been solely to maintain it as an "empty space" (free of all hostile forces) at minimal cost, although this objective has failed doubly, as the Taliban have re-emerged in strength and the occupation is turning out to be very costly.[107]

According to Giustozzi's interpretation, after 2001 there was a key political dilemma in Afghanistan between the short and the long term. In the long term, the consolidation of the State and peace demanded the establishment of a central authority which would be in control of force and the creation of institutions. Nonetheless, these processes could cause conflicts in the short term with the warlords and the strong men in the country. This dilema led to the division of the international community into two major orientations. The first claimed that the warlords, mercenaries and all those involved in violence over the previous two decades ought to be excluded from political life and from power, in order for the Afghan State to have a new start. This was the opinion of many UN officials, particularly those with ample experience in the country, as well as that of human rights workers and some social scientists. On the other hand, the second stance felt that the priority was maintaining stability in the short term and that this could only be achieved in concert with the non-State actors rejected by the other position. Those who defended the second position were mainly diplomats and experts in international relations. For many of them the "collaboration" was merely a stop-gap measure, and not a long-term strategic option, without paying much attention to the matter of how to establish a basis so that the Afghans could construct institutions in the medium and long term, which is something qualitatively different and opposed to the mere occupation of the State by the various factions and the struggles for areas of influence and the division of booty.

The United States policies prioritized short-term stability over long-term political change. As a result, this has been the prevailing focus since 2001. Institutionalization and the 'primitive accumulation of power'[108] by the central authority have been pushed aside. The short-term risk of destabilization has been avoided by not offending the warlords and other potential losers. The problem is that the initial commitments with the warlords have impeded a rapid consolidation of an effective State and left large parts of the rural areas in disarray. All of this has left the door

---

[107]HEROLD, Marc W, *Afghanistan as an Empty Space ...*, cit., p. 2 (where the quote is taken from) and *passim*.

[108]An expression which paraphrases economic jargón. It is used by Giustozzi, following among others, COHEN, Youssef et al., "The Paradoxical Nature of State-Making: The Violent Creation of Order," *The American Political Science Review* 75, no. 4 (December 1981): 901–10, and TILLY, Charles, *Coercion, Capital and European States, AD 990-1990* (Cambridge MA: Blackwell, 1990). It corresponds, in part at least, with what Weber called the establishment of the monopoly of violence. Giustozzi further considers that what would be involved in the phase of 'primitive accumulation of power' would be the monopoly of violence on a large scale. To the contrary, the control over violence on a small scale would be the concern of a mature State.

(GIUSTOZZI, Antonio, *Empires of Mud. War and Warlords in Afghanistan*, cit., p. 10).

open to a revival in the long run of the war cycle, although current threats do not stem from the former military class or older warlords, but rather from other factors.

In the second half of the post-Taliban decade the veteran warlords (Dostum, Ismail Khan, the remaining Shura-i Nezar) were politically weakened, because the effects of international intervention had impeded consolidation of the incipient processes of institutionalization of their governments (policies). They had degenerated into mafias and gangsterism. On the other hand, the advanced age of the leaders, the accumulated riches and the sinecure obtained from the Kabul Government kept them from radical options.

That said, a new cycle of war began in 2006 caused by new factors of destabilization. Principally, the increase in insurgency and the change in the mood of the population who became progressively distanced from the government, even voluntarily supporting the insurgents.[109]

Furthermore, in Afghanistan, the security reconstruction failed, normally considered an immediate priority in the process of state recuperation. Neither international forces, nor the new Afghan security forces (the Afghan National Army and the Afghan National Police) have managed to re-establish law and order or to offer security and protection to the population against the warlords and the Taliban. Among other things, many former members of the armed groups were incorporated into the police, the army and the security services while maintaining de facto their ties to the militias.

The economic reconstruction has been inexistent. The country suffers high levels of unemployment. The reason for this is that the real economy is stuck. The country's economy depends on imports, which apart from creating a deficit, impedes the creation of employment, and even destroys it. On the other hand, the daily struggle for survival generates a large submerged economy which is estimated to represent between 80% and 90% of the entire economy of the country.[110]

Certainly, some sectors of the economy have grown greatly since 2001, in particular the cellphone industry, banking services and construction. The first two, however, generate relatively little employment. And as far as construction, a significant portion of the jobs are taken by workers coming from Pakistan due to the lack of qualified labour in Afghanistan. Furthermore, the construction boom began to ebb in 2008. Agriculture and livestock, upon which 80% of the population depend, have had very unstable results over the past years. In practice, the country's economy depends on international intervention and on the informal economy (especially the production and trafficking of drugs, along with other types of traffic and contraband).[111]

---

[109]GIUSTOZZI, Antonio, *Empires of Mud. War and Warlords in Afghanistan*, cit., "Conclusion", pp. 297–305 passim.

[110]HEROLD, Marc W, *Afghanistan as an Empty Space . . .*, cit., Section 2, pp. 27–38.

[111]GIUSTOZZI, Antonio, "Afghanistan: getting worse before getting better?", cit., pp. 1–3.

Nearly the entire budget of the country comes from abroad. Furthermore, a large part of it is earmarked for maintenance and security and a good portion is lost to corruption. The Karzai regime depended almost entirely on external aid; not only military but also in terms of its financing. The regime seems to have accepted this economic dependence as a long term reality, along with the growing role played by drug trafficking in the country's economy. At the same time, as throughout most of Afghanistan's history, no attempt has been made to develop the country's industrial framework.[112]

Furthermore, the presence of numerous international actors: UN Agencies and other international organizations, as well as those of different governments, NGOs, foreign troops, has a strong impact on the economy, not only because of the expenses and payments by these agencies, but also because it seriously distorts the economy with regard to specialized employment and salary expectations. International agencies hire the few Afghans who are qualified, which has led to a salary rise which makes it impossible for the Afghan government to hire qualified people.[113]

Opium cultivation and heroin production is one of the few sectors which can guarantee a living for the impoverished rural population of Afghanistan. It is currently estimated that the value of the drug economy has reached double the amount of western aid, including development assistance and financing for reconstruction.[114]

The situation described generated enormous disappointment among the population and has led to the resurgence of the Taliban, mainly due to the failure of political and economic reconstruction. In this sense, Giustozzi claims that the arbitrary nature of the post-2001 government was one of the main advantages of the neo-Taliban, whose social base lay in the communities in conflict with the local governments and the police. Another key factor was Pakistan's involvement in supporting the insurgency in Afghanistan. As a result of all this, in 2007 the neo-Taliban represented a serious threat.[115] Since then the situation in the country has only deteriorated.

The considerable collateral damage caused by the conflict has led to a rejection of the international forces and a favouring of the insurgency. "A weak government, combined with civilian casualties from combat operations, is the main factor allowing the Taliban and Hezb-e-Islami to increase their reservoir of support and numbers". The increase in military pressure did not aid in the fight against the insurgency, and may have had the opposite effect. This would be the main argument

---

[112]GIUSTOZZI, Antonio, "Afghanistan: The Patrimonial Trap and the Dream of Institution-Building", cit., p. 79, and on the lack of economic modernization in the past, pp. 72–5.

[113]GHANI, Ashraf and LOCKHART, Clare, *Fixing Failed States* (Oxford: Oxford University Press, 2008), p. 100.

[114]DUFFIELD, Mark, *Development, Security and Unending War . . .* , cit., p. 183.

[115]GIUSTOZZI, Antonio, *Koran, Kalashnikov and Laptop: The Neo-Taliban Insurgency in Afghanistan 2002-2007* (New York: Columbia University Press, 2007), pp. 11–79.

in favour of a political solution to the conflict and the need to negotiate with the Taliban (and the *Hebz-e-Islami*) in a joint effort by two members of UNAMA (Masadykov and Page) as well as a researcher at the Crisis States Research Centre.[116]

4.7 On the Necessary Recuperation of Politics in the Face of the Violence and the Responsibilities of the International Community

It is certain that the Afghan conflict demands a political solution, but not merely in the sense of a mere division of power. This would correspond to a patrimonialist vision of politics as applied broadly since 2001 with dreadful results. If the only novelty lay in including the Taliban in this division of power, from which they were excluded in 2001, I do not believe that the long-term results would be much better, though in the short term the situation would be stabilized. Again there is the problem between the short and the long term.

Efforts must be directed, then, toward reconciliation and recovering the political arena in the most noble sense of the term, as a means for a society to act in harmony. This will of course require the incorporation of all sectors of the society in question, as well as the Taliban and the former Mujahedeen and warlords, but without anyone having recourse to the means which would allow them to employ violence to alter the results of the political game.

One reason that, following 2001, there was a failure of both the disarmament process and the transformation of political-military groups into unarmed political actors, was that these political-military groups did not wish to enter the competition in the political arena under equal conditions with the unarmed groups, against whom they had been at a disadvantage, particularly in the urban areas where some of these unarmed political groups had a long tradition and strong ties to the population. For this reason, intimidations and threats carried out by mercenaries linked to political-military groups were aimed specifically at the civilian parties rather than rival militias.[117]

The rich distinction between political power and violence, as underlined by Hannah Arendt, may cast light on the significance of the current situation in Afghanistan from the political theory and philosophy point of view, as well as on possible (and no doubt difficult) ways out.

Arendt points out that there are two ways to understand political power. The first of these has prevailed, and has had an extraordinary influence on western political

---

[116] MASADYKOV, Talatbek, GIUSTOZZI, Antonio and PAGE, James Michael, "Negotiating with the Taliban: . . . ", cit., passim. Quote from p. 8. In the same line, SUHRKE, Astri, "Afganistán: ¿negociar con los talibanes?", *Política Exterior*, no. 129 (mayo–junio 2009): 45–70.

[117] GIUSTOZZI, Antonio, "Afghanistan: Political Parties or militia Fronts?", cit., p. 197.

tradition, particularly in the line of authors from Bodino and Hobbes to Weber. It conceives power as the dominion of man over other men. From this perspective, the essence of power consists of ruling and being obeyed, known as the "efficiency of rule". In keeping with this conception, power can not be distinguished from violence, or may be but with great difficulty. The words of Mao Tse Tung, according to which "power comes from the barrel of a gun" would be the crudest and most extreme expression of this version of power.

There is, however, another conception of political power, which is that defended by Arendt, as the ability of a group or people to act together. "Power springs up whenever people get together and act in concert". Power is the condition which allows men to live together and pursue common ends. Understood thus, "[p]ower is never the property of an individual; it belongs to a group and remains in existence only so long as the group keeps together". In keeping with this understanding, power is based on the backing of the people, on the force of opinion.

This vision of political power appears intimately linked to the idea of citizenship in the classic Greek sense, according to which – as Adela Cortina states – the citizen is "one who is concerned with *public issues* and is not content with minding only his private affairs; but is also one who understands that *deliberation* is the best way to address them, more than violence, more than imposition; even more than voting, which is the final recourse, when the force of the word has already been deployed". These are the bases upon which the civic republican tradition rests "which understands politics not as the moment of the legitimization of violence, but as the overcoming of violence through communication".[118]

Power and violence -as Arendt continues- although distinct phenomena, normally appear together. In these cases, power is the primary, or predominant factor, which serves as a guide, while violence has a purely instrumental character, as a last resort to maintain power structures intact (against delinquents, or internal rebels or against outside enemies).

Nonetheless, there are times that violence is completely separate from political power. When this happens, it is very difficult to invert the situation, given that violence (as a pure means) becomes the end and tends to destroy any power.[119]

One could say that the latter is what happened in Afghanistan. At first the violence was guided by properly political objectives, in Arendt's sense, such as an expression of a popular resistance movement against the Soviet invasion (in the case of the mujahedin), against the warlords' violence (in the case of the Taliban), against the foreign occupation (in the case of the insurgency which re-emerged after 2003). Nonetheless, the violence of these groups quickly split from any common purpose, becoming progressively an end in itself for those who took exclusive advantage of it. One could add to this the violence inherent in the very presence and actions of the international forces.

---

[118]CORTINA, Adela, *Ciudadanos del mundo. Hacia una teoría de la ciudadanía*, 2ª ed. (Madrid: Alianza Editorial, 2005, 4ª reimpresión), pp. 44–5.

[119]ARENDT, Hannah, *Crises of the Republic* (San Diego/New York: HBJ, 1972), pp. 134–155.

In these circumstances where the violence tends to destroy any power whatsoever the only way out seems to be the attempt to reconstruct, slowly and with difficulty, the ability to act together.

This, in the first place, refers to the principle actors of present day international politics: The United States, The European Union, NATO and the United Nations. It is necessary to overcome what Habermas called the "rupture of the West",[120] which was motivated principally by the crisis in Iraq, but in which a part had been played, although secondary, by the frictions on the Afghan stage among the United States on one hand and NATO on the other, due to the former's refusal to support the disarmament of the warlords' militias.[121]

A regional diplomatic initiative is similarly needed, to promote dialogue between the states of the region: Pakistan, which is undoubtedly the key player, but also India, Iran and the five republics of Central Asia. Understanding must begin to relieve the multiple tensions among them and obtain an agreement not to interfere in the Afghan conflict. Thus, for example, the conflict among India and Pakistan is what guides the attitudes of both countries regarding Afghanistan. The reason why both attempt to influence Afghanistan is in order to seek solutions to their own internal insecurities. This means that "as long as the core insecurities within the South Asia Regional Security Complex are not resolved, negative influences may continue to hamper stabilization efforts in Afghanistan".[122]

Internally, the national construction of Afghanistan demanded a change in the current situation, in other words, to deactivate the violent parties and thus leave space to exercise political power. Both objectives, sides of the same coin, involved difficulties, particularly the former.

In order to deprive the armed groups of their "dominion", there was first a need for the disarmament, demobilization and reintegration (known as DDR), to be complete and effective. Although there were many delays and difficulties, the United Nations finally managed, in 2003–2004 to implement a disarmament program which managed to collect heavy arms from the warlords and to disarm a good number of militiamen, which was well received by the population, although hundreds of groups of armed men continued to work around the country. In the medium term the effective results of this partial disarmament were not very encouraging. The success of the disarmament programs depended on achieving sustainable reintegration for ex-combatants into ordinary civil life. For this, it was necessary to be able to ensure sustainable income. Thus reintegration was tightly linked to the creation of employment and to economic development. Furthermore, these same conditions were indispensable in order to cut the source of new recruitment.

---

[120]HABERMAS, Jürgen, *The Divided West* (Cambridge: Polity Press, 2006).

[121]RASHID, Ahmed, *Descent into Chaos . . .* , cit., pp. 202 y 209–11.

[122]TADJBAKHSH, Shahrbanou, *"South Asia and Afghanistan: The Robust India-Pakistan Rivalry"* (Oslo: Peace Research Institute Oslo, 2001), p. V y passim.

This led to the need for an economic development program for the region as a whole.[123] Particular attention needed to be given to education and the creation of employment in the border regions between Afghanistan, Pakistan and Central Asia.

It is often noted that in order to pacify Afghanistan and impede the Talibanization of the border regions with Pakistan, which would affect not only the tribal agencies (FATA), but also certain zones in the Northwestern Border Province, a reform of the FATA would be necessary in order to put an end to their political and economic isolation, inherited from the colonial era, by integrating them into national policies and state administration. In addition, the ICG Report from 2006[124] proposed integrating the FATA within the Northwestern Border Province, developing its natural resources and encouraging agriculture. Only thus would it be possible to disarm the militant Islamists, close the training camps, persecute those responsible for the murder of civilians and civil servants and open the FATA to the press and to human rights groups, as this group suggests. Along with this, it would be necessary to clearly recognize the Durand Line as an international border for both Afghanistan and Pakistan.[125] This would relieve Pakistan's concern for its security, which has coloured all of its international policies since its creation in 1947, and facilitate its support for a political agreement with Afghanistan.

Similarly, an issue which is often raised is the importance of addressing poverty and the deficiencies of the Afghan agricultural system, in order to supply the rural population with alternative means of living other than opium production. For this, the substitute crops must be viable, sustainable for the medium and long term and able to effectively guarantee the sustenance of the communities. Along with this, it is important to address the problem of reconstructing the deteriorated, or utterly destroyed irrigation and transport infrastructures.

On the other hand, national reconstruction requires the recuperation of the exercise of genuine political power, in Arendt's sense, which is the Afghan people's ability to act in concert through their political institutions.

Faced with the dominant reality until now (the commitments and consequent divvying up of power with the warlords and drug traffickers), an attempt should be made to introduce those who have the ability to act in concert which are those who can contribute most positively to national reconstruction and the consolidation of peace. In this, an important role can be played by the political reformist leaders and true democrats, who in the Afghan political context have always been silenced and made invisible, in Malalai Joya's words.[126] It is also important to leave space and establish channels for citizen participation (under secure conditions) for the majority

[123]Rashid mentions the need for a new "Marshall Plan" for the region of Central Asia and the South (RASHID, Ahmed, *Descent into Chaos . . .*, cit., p. XLIV).

[124]Internacional Crisis Group, *Pakistan's tribal areas: appeasing the militants*, 11 December 2006.

[125]ESPINOSA, Ángeles, "Las áreas tribales pakistaníes y la lucha contra el terrorismo", cit., pp. 61–2.

[126]Afghan Member of the Parliament who faced the warlords at the Constitutional Loya Jirga and since that moment has been persecuted.

of the citizens, who are peace-loving and disarmed, who for decades suffered the violence of various groups and have developed the ability for resistance, solidarity and acting together through community structures and civil society. As Adela Cortina points out, civil society is at the present the best school of citizenship, as in civil society groups, established freely and spontaneously, people learn to participate or show interest in public matters.[127] Within this peaceful and disarmed citizenry who can contribute positively to pacification and reconstruction of the country, women play an outstanding role. RAWA (Revolutionary Association of the Women of Afghanistan) was notable for its peaceful resistance to the Soviet invasion, against the warlords and the Islamists as well as the Taliban, for their humanitarian work in refugee camps in Pakistan and Afghanistan and for having created extensive solidarity networks which were highly consolidated in the territory. Nonetheless the attitudes and policies with regards to women in the groups currently in power in Afghanistan make it difficult for women to have a real influence on political life.[128]

To summarize, the idea is to construct a democracy with a genuinely popular base, in which the Afghan author Jawad placed his hopes for peace in his country in 1992.[129] Nearly two decades later, it remains a pending challenge.

As I have attempted to show throughout this paper, the external interferences, albeit interacting with internal dynamics, have triggered the process of replacing politics with violence. Once the Pandora's Box has been opened, it must be the Afghans themselves who must slowly and painfully reconstruct a pacific political coexistence, which will replace the decades of violence suffered by the country.

Having said that, this should not mean one should forget the responsibility of the international community in this matter. As we have seen, current military intervention has not contributed to the pacification of the country, but has instead had the effect of increasing the conflict. Following this, the international community must not repeat the same errors of the 1990s; in other words, they must not ignore and abandon Afghanistan following the withdrawal of the international (North American and NATO) troops. On the contrary, the international community must assume a serious, prolonged commitment to the establishment and consolidation of peace in Afghanistan during the period of the progressive withdrawal of foreign troops, as well as after this is completed in 2014.

The lesson which may be taken from the experience of the 1990s is that although the troop withdrawal will effectively end the international conflict, it is probable, given the profound destructuring of the country politically and economically, that the conflict will persist at a local level. This is in itself serious because the suffering of a people and of the individuals of which it is made is always cause for concern.

---

[127]CORTINA, Adela, *Ciudadanos del mundo* ..., cit., pp. 36–7 y capítulo V, "Ciudadanía civil. Universalizar la aristocracia", pp. 133–75.

[128]MANN, Carol, *Femmes afghanes en guerre*, cit., pp. 131ss.

[129]PHILLIPS, Alan, "Preface", in a JAWAD, Nassim, *Afghanistan. A Nation of Minorities*, cit., p. 5.

Furthermore however, sooner or later it could have global repercussions. In the global era, collective security, the security of all, can only be the result of respect for the rights of all and for each human being in all parts of the world.

The international community, which was prepared to maintain armed intervention in Afghanistan for over a decade (from 2001 to 2014 when withdrawal of foreign troops is expected), must be ready to remain in the country through a UN peacekeeping mission, which could be the UNAMA with a new mandate separating it from the ISAF, or a new mission authorized for this purpose by the Security Council, for the time necessary and to dedicate the necessary resources in order to establish a base for the consolidation of a lasting peace in Afghanistan.

Chapter 5
Somalia: From the Errors of Colonialism to the Horrors of War

Ana-Paz Garibo-Peyró

Abstract Chapter 5 analyses the reality of the Republic of Somalia, a country which is currently facing the worst humanitarian crisis in history. The first descriptive part of the article shows how the current human catastrophe devastating the Somali people is the result of the incomprehensible de-composition of one of the most homogenous peoples in Africa, not only ethnically speaking, but also from a linguistic, religious and cultural point of view. This de-composition, which has led to Somalia being known as the "State without State", is undoubtedly rooted in the errors and horrors suffered, in the first place, during the successive processes of colonization and de-colonisation, but also during the short history of the State's independence, which instead brought about the road towards the civil war which finally broke out more than 20 years ago. Taking the reality of Somalia as a limit situation, the second prospective part of the chapter reflects upon the demand for a change of paradigm concerning fundamental questions of International Law, such as the need to transform national security into human security, the step from humanitarian intervention to the responsibility to protect, and the urgent need to examine the bases in greater detail in order to reconstruct failed States.

A.-P. Garibo-Peyró (✉)
Philosophy of Law, University of Valencia, Valencia, Spain
e-mail: A.Paz.Garibo@uv.es

J. Ballesteros et al. (eds.), *Globalization and Human Rights*, Ius Gentium: Comparative Perspectives on Law and Justice 13, DOI 10.1007/978-94-007-4020-4_5,

5.1 Introduction: The Gravest Humanitarian Crisis in the World

Currently, Somalia appears at the top of the list in the international Failed States Index.[1] The Failed States phenomenon reflects the actual international scene and the concerns which political analysts consider a threat to international security.[2] Somalia's situation seems to be especially critical in this regard. In the opinion of ROYO ASPA, Somalia represents the paradigm of a Failed State.[3] ROTBERG has studied this concept, establishing a classification of stages of failure, with the "collapsed state" as the most extreme and unusual degree of failure. In this author's view, Somalia is the prototype of the collapsed state; indeed, it is the only one that remains in a "collapsed" situation while other countries like Sierra Leona, for instance, were collapsed at a certain moment but were able to recover from the problem.[4] More recently, GUGLIELMO underlined the fact that more than a failed state, Somalia, is a fragmented state with several smaller territorial realities, where thousands of political, clan-family or personal organisations vie to settle policies regarding territorial control and, also, international relations.[5] Particularly significant is the title of an early article published by MENKHAUS,[6] in 1995: "Somalia, the Stateless State". Perhaps this is the qualification that describes the reality of this country most accurately because since the fall of the dictator, Siad Barre, in 1991, Somalia has been going through a long, complicated and violent, armed conflict, and has no central government with sufficient power, but rather many different local authorities, of differing natures, in confrontation with each other.

It is hardly surprising that Somalia is currently living the gravest humanitarian crisis in the world. Two generations of children have known nothing but war. Today, approximately one third of the entire population relies on humanitarian aid to survive because of the devastating effects of war and disastrous crops. In the latest confrontation more than 9,000 people have lost their lives, and 1.1 million have had to abandon their homes. Besides all these disasters, Somalis are suffering one of the droughts that cyclically devastate this region of the earth.

Far from resolving itself, the political crisis is getting steadily worse. Violence and insecurity hinder the possibility of accessing humanitarian aid. In January

---

[1]See "The Failed States Index 2010" *Foreign Policy,* available at http://www.foreignpolicy.com/articles/2010/06/21/2010_failed_states_index_interactive_map_and_rankings.

[2]FERNÁNDEZ, E., *¿Estados fallidos o Estados en crisis?* (Granada: Comares, 2009).

[3]ROYO ASPA, J.M., "Las sucesivas crisis de Somalia", *Fride*, Review, September 2007, p. 1.

[4]ROTBERG, R. I., "The Failure and Collapse of Nation Sates: Breakdown, Prevention, and Repair", in *When States Fail: Causes and Consequences*, ed. Robert I. Rotberg (Princeton/Oxford: Princeton University Press, 2004).

[5]GUGLIELMO, M., *Somalia, le ragioni storiche del conflitto* (Torrazza Coste: Altravista, 2008), p. 169.

[6]MENKHAUS, K., PRENDERGAST, J., *Somalia: The Stateless State*, Africa Report 40:3, (1995:May/June), p. 22.

2009, 24 aid-workers lost their lives in Somalia.[7] More than half of the Somali population is under age.[8] If we consider that children are one of the most fragile and vulnerable human groups, it is evident that they suffer, in a very tragic way, the worst consequences of their country's situation.

The question is: how has the situation degenerated to this point? How is it possible for an armed conflict to last so long (more than 20 years), despite the numerous attempts by the international community to put a stop to it? Why is the war intensifying day by day? What solutions can expert international analysts supply?

It is impossible to answer all of these questions without trying to understand the social, cultural and territorial features of Somalia and addressing the historical root causes of the crisis, in depth. Given the limited scope of this work, it will not be possible to undertake in full such an ambitious aim but I will attempt to provide enough historical information to allow us, afterwards, to reflect on the different interventions and solutions that both the international community and the Somali community have been trying for years.

5.2 Historic Somalia: Somali People

Somali peoples have inhabited the Horn of Africa since time immemorial, occupying the territory that nowadays includes the Republic of Djibouti, the current Republic of Somalia, North Eastern Kenya and the Ogaden, that is, the easternmost region of Ethiopia. Somali land was bounded by the Red Sea and the Gulf of Aden to the north, by Kenya to the south, by the Indian Ocean to the east and by the Ethiopian empire to the west. Almost the entire country is made up of desert plains and plateaus, except the Golis Mountain range in the Northwest, whose highest peak, Surud Ad, reaches only 2,408 m. The northern region is dry, hot and inhospitable; it's known as "the *guban*", a Somali word that means "burnt". Rainfall is too low (and droughts cyclical), so the vegetation, lower woods and steppes, make it suitable only for breeding sheep, goats and camels. The country has only two rivers that cross the southern region from west to east; both have their sources in the highlands of Ethiopia: the Jubba river flows into the Indian Ocean, in the area around Kisimayo city, and the Shabeelle river flows into the marshes of Hawai, near the little town of Brava.

---

[7] Re: the recent worsening of the humanitarian crisis, see Steve BLOOMFIELD's article, published in the Scottish newspaper The Sunday Herald, 31 January, 2009. See also, CORNWELL, R., "After peace comes the storm: Somalia's relentless crisis", *Fride,* Comment, August 2008.

[8] Re: the number of inhabitants in Somalia, there is no exact data and the figures vary depending on the source of information. The Child Soldiers Global Report 2008, published by Coalition to stop the use of child soldiers, p. 305, talks of 8.2 million inhabitants, of which 4.2 are under 18 years of age (the report is available at http://www.child-soldiers.org/library/global-reports). Amnesty International's 2008 report talks of 8.8 million Somalis, see p. 347. This last report is available at http://thereport.amnesty.org/document/40.

The valleys formed between both rivers are the only fertile land in the country. Indeed, the temperatures are more clement although there are ostensible drops in the dry season and the rainfall is much higher than in the North.[9]

This entire region forms a well-defined geographic and ethnic unity that has been considered, over the years, as a natural base on which to build a sovereign state.[10] Certainly, Somalis are one of the most homogeneous nations of the continent. They are the exception in black Africa because all of them belong to the same ethnicity, share a language of oral tradition (Somali), a culture and a way of life and, most importantly, profess the same religion, Islam.

Although there are certainly some differences between them, they have common features that make them a particular people of Africa. They belong to the ethnic group of the Hamites, characterized by their stylised beauty. The closest tribes to them are the Ethiopian and the Eritrean from the East and the North, the Afar who live alongside them in Djibouti and, lastly, the Oromo and Bantu from Kenya.[11] Their facial features suggest a secular contact with Arabic people too. Indeed, Islam arrived in Somalia in the eighth century.[12] For centuries, Somalis have also lived with minorities of Bantu and Swahili origin.

In 1965, the British anthropologist I.M. Lewis estimated that only an eighth of Somalis were sedentary farmers in the rivers region in the South, nearly all of them belonging to the Digil and Dahanweyn clans.[13] Most Somalis are nomadic, belonging to other clans that work with animals and travel hundreds of kilometres a year, moving from place to place in search of water and pasture land. Nomadic Somalis travel around the North and centre of the country and, also, around the south of the Jubba on the border with Kenya.[14] However, Guglielmo and Ruiz-Giménez agree in estimating the current nomadic population to be approximately 60% of the total, while the remaining 40% constitute the sedentary population working as farmers, artisans and traders.[15]

---

[9]Re: geographical and orographic data on Somalia, see the Report on the Democratic Republic of Somali published by the Spanish Ministry of Foreign Affairs and Co-operation in February, 2008, available at http://www.maec.es/SiteCollectionDocuments/Monografias/Somalia.pdf. See also, GUGLIEMO, M., *Somalia. Le ragioni storiche del conflitto* (Torraza Coste: Edizioni Altavista, 2008), pp. 25 and 26.

[10]LEWIS, I. M., *A modern History of the Somali*, 4th ed. (Oxford: James Currey; Athens: Ohio University Press, 2002), p. 1.

[11]Vid ibid, p. 4 and following.

[12]Re: reasons for the dizzying spread of Islam among neighbouring peoples, see MERNISSI, F., *El miedo a la modernidad, Islam y democracia,* in Eastern and Mediterranean publication, 1992. One must remember that in the year 711, Islam arrived in southern Spain, at the opposite point of the Mediterranean.

[13]I refer to the first edition of the work by the cited author.

[14]LEWIS, I.M., op. cit., p. 7 and following.

[15]RUIZ-GIMÉNEZ ARRIETA, I., Las "buenas intenciones", *Intervención Humanitaria en África* (Barcelona: Icaria, 2003), p. 62; GUGLIELMO, M., op. cit. p. 26.

The Somali language, of oral tradition,[16] contains a considerable amount of terms with Arabic roots; indeed, the Arabic language is so widely spoken by Somalis that it could be considered a second language.

Islam and many other elements from Arabic culture are deeply linked to the Somali people but with their own peculiarities: so Islam and the Islamic law, the Shari'a, merged with consuetudinary Somali law giving rise to the Xeer, their own version of the Islamic law which the Islamic Courts apply. The Xeer is less rigid than the Shari'a.[17]

The Somali social structure is centred around so-called "clan-families", which, in turn, are subdivided into clans, sub-clans, lineages and so on. But the most primitive division of this people is that which distinguishes between "Samaal" (or Somalis proper) and "Saab". Somali people claim a noble ancestry which dates back to Mohammed; this ancestry would join all the clan-families in one unique genealogical system through the maternal line (except the Darod linked by the paternal line). In this way, the Dir, Isaaq, Hawiye and Daron descend from a common ancestor called Samaal, while the Digil and Rahanwyn believe that their ancestry dates back to their common ancestor called Saab. According to Lewis, the custom of calling both groups Somalis is an error similar to the one we make calling all the inhabitants of the British Isles English.[18] Saab clan-families (Digil and Rahanweyn) are sedentary and work mainly as farmers and, as we have just said, they live in the southern valleys, between the Jubba and the Shabeelle rivers. They barely exceed half a million people. Samaal clan-families, much more numerous, (Dir, Isaaq, Hawiye and Darod) are principally nomadic shepherds distributed across the Somali country. Certainly, many clear economic and cultural differences exist between Saabs and Samaals. As Lewis has underlined, the nomadic Samaal have traditionally looked down upon and disdained the sedentary Saab, which has created a sense of resentment and isolation between them. However, the abyss between them is not as deep as it might seem at first sight, since the Somali clan is exogenous so that many Saab come from the pastoral North. Moreover, both groups speak the two dialects of Somali and their economies and life-styles are complementary.[19]

In short, the six great families of clans, nomadic and sedentary, have developed primary economic systems for centuries, the only possible way in such a hostile environment with no natural resources. It is important to stress the fact that their communication and exchange networks, economic and cultural, have always been much more intense with the Arabic world than with their Ethiopian or Kenyan neighbours, most of whom are Christians. Since the tenth century, they have also

[16]The Somali language was not transcribed until 1972, during the first years of General Barre's dictatorship, as we will see later.

[17]Cfr. RUIZ-GIMÉNEZ ARRIETA, I., *op. cit.*, p. 63.

[18]LEWIS, I.M., *op.cit.*, p. 6.

[19]LEWIS, I.M., *op. cit.*, p. 14.

carried on trade relations with India and China. They did not have any contact with European imperialism until the nineteenth century.[20]

However, long before the European powers arrived in Somalia, this country already had its own political system. Clan-families, the highest level of identification and unity, as we already know, are so many and so wide-spread that they could never act as a political unity. However, that is what they have done. Every clan is organised in accordance with the "shepherd democracy" or "democracy under the tree", which consists of the Assembly (shir) of all the men of the clan (so it excludes women) for decision-making about resources access and trade routes control. Such assemblies are ruled by the Xeer. On the other hand, inside the clan system, we find the so-called "segmentary lineage", where political loyalties are decided according to genealogical proximity. This "segmentary lineage" lies in alliances between clan-families in order to secure resources and, also, to decide on access to the resources (most of all, water and pasture) and on the control of trade routes. Unfortunately, the alliances have been historically fragile, fickle and unstable. The formula of clan-alliances plays a seemingly contradictory role because they get stronger in the face of external threats but, when these threats disappear, they have internal confrontations. For generations, the most important political and social factor has been clan membership, so much so that loyalty to the clan has even been able to undermine the common sense of belonging to the same nation.[21]

The social, economic, political and cultural overview concisely described above would obviously change when the Somali people, like the rest of the African continent, became the victim of European imperialistic ambitions.

5.2.1 Sharing Out Somalia

It is a commonplace to consider that it was during the Berlin Conference (that took place between November 25th 1884 and February 26th 1885, and was fostered by Bismarck) when the European colonial powers shared out the African continent.[22] But, according to WESSELING, things were more complicated. At Berlin, the sovereignty question of the different regions of Africa was not a matter included on the agenda, as the Chancellor reminded the attendants in the opening session. The items on the agenda were the following[23]:

1. Freedom of trade in the basin and mouth of the Congo

---

[20] RUIZ-GIMÉNEZ ARRIETA, I., *op. cit.*, p. 63.

[21] Re: pre-colonial Somali political system, see: MAKINDA, S., "Somalia", Adelphi Papers no. 32, 1992, p. 25 and following; and RUIZ-GIMÉNEZ ARRIETA, I., op. cit., p. 62 and p. 63.

[22] See ROYO ASPA, J. M., "Las sucesivas crisis de Somalia", *Fride*, review, September 2007, p. 3. Available at: http://www.fride.org/publicacion/255/las-sucesivas-crisis-de-somalia.

[23] WESSELING, pp. 147–152.

2. Freedom of navigation in the Congo and the Niger based on the same principles as applied for the Danube.
3. The definition of the formalities to be observed when taking possession of the new territory on the African coast.

Obviously, the last point was the most important: "the rules for the occupation of new territories in the African coast",[24] as Wesseling has said, but what about the hinterland? The sovereignty question had been excluded from the Conference and, nevertheless, this was indeed the most interesting point for all the colonial powers, and it was just this issue which was debated along the corridors. It is important to state exactly what the delegates actually agreed, "since this item on the conference agenda has become a source of myth and confusion".[25] The frontiers along the coasts were easy to delimit (from latitude to latitude), the problem was to work out where the coast ended and where the hinterland began, and above all: what to do with the hinterland? Considering all the sovereignty aspirations and measuring the territories was equivalent to making partitions of Africa, and that was not the purpose of the Conference.

Except for the question of the Congo Free State (arbitrarily recognised as a corporate state, privately controlled by Leopold II of Belgium), resolved at the aforementioned Berlin Conference amongst the colonial powers that had trade interests in the region, the partition of Africa, as a general rule, "was ultimately a matter of bilateral diplomacy".[26] In the case of Somalia, the partition was carried out as follows:

France had created the port town of Obock, in 1862, on the coast of the current Republic of Djibouti, the population of which was mainly Somali. For its part, Great Britain had already been in Aden in the south of Yemen since 1839 and took its supplies from the opposite coast of Somalia. In 1884, more by chance than deliberately,[27] the British established a protectorate inhabited by Somali people in the north of that region and, in 1887, placed under protectorate status the Sudanese coast in the north of Eritrea.

The border between French and British Somalia was established in 1888. Djibouti became the capital of the French colony. The Somali country under the British Empire would come to be known as Somaliland.

In 1887, Italy got a protectorate from the Sultan of Zanzibar that stretched from Kisimayo city, situated in the South of the current Republic of Somalia, next to the

[24]WESSELING, p. 151 in the Spanish edition.

[25]WESSELING, *op. cit.*, p. 151 in the Spanish edition.

[26]WESSELING, *op. cit.*, p. 159 in the Spanish edition.

[27]The Mahdist revolt in Sudan required a rapid retreat by the Egyptian troops from the Horn of Africa, threatening an ensuing power void. The British had to supply their troops through the Red Sea and Aden. On the other hand, they were worried by the imperialist ambitions of the other powers in the area (France, Italy and Ethiopia), so the British government established several protectorate treaties with local leaders of Somali clans in the North. Cfr. Somaliland: Democratisation and its discontents, International Crisis Group, Africa Report no. 66, 28 July 2003.

estuary of the Jubba River, to Cape Guardafui, the easternmost cape in the African continent just at the tip of the Horn of Africa. Then, Great Britain and Italy held diplomatic negotiations in which they shared out their respective influence zones in Somalia: in two treaties, signed on March 24th and April 15th, they agreed that Italy would take supremacy over Somalia, establishing a border with the British at the Jubba River. Furthermore, Italy undertook to keep away from the Nile River in its colonial aspirations.

Under the reign of Menelik II, Ethiopia was an expanding empire that started its advance towards the Red Sea. It obtained control of the Somali region of Ogaden through an agreement with Great Britain in 1897.

On the other hand, the frontier between the Italian and Great Britain territory in the South of Somalia was defined by a treaty signed by both powers on July 15th, 1911. Thus, part of the southern Somali territory remained under Great Britain (in the North-eastern district of Kenya).

In brief, this was the way in which Somali land was divided into a five-piece mosaic occupied not only by European powers (France, Great Britain and Italy), but by a continental empire as well (Ethiopia).

At this point, it would be interesting to consider the particular features of the imperialism of every power that occupied Somalia, in order to gain a better understanding of the consequences of the different forms of colonisation and their impact on the present situation of this country. Thus, we will briefly look at the characteristics of the Imperialism of Great Britain, as well as that of France, Italy and Ethiopia.

5.2.2 The Imperial Powers

5.2.2.1 The British Empire

In the nineteenth century, British society was the most dynamic in Europe. In continuous growth, it also enjoyed a stable political system based on the alternation of power between Tories and Whigs. All of this brought a great increase in prosperity and an expanding economy. Its colonial, trade and maritime supremacy was evident during the French Revolution Ward and, also, during the Napoleonic Wars. Indeed, the victories at Trafalgar and Aboukir put a brake on France's imperialist and expansionist ambitions. At the beginning of the nineteenth century, the British Empire was already impressive[28] and it continued to spread.

---

[28]"It comprised Canada and a number of important West Indian possessions; trading posts in West and East Africa and the great white colony in the Cape; British India, the pearl in the British crown, in Asia, together with Malaysia; in Oceania, the white colonies of Australia and New Zealand and, finally, a great number of naval sea bases along the different sea routes. In the further course of the nineteenth century British expansion continued apace, if possible by

The British overseas expansion was different; its empire was unique because it was not merely a political expansion like the French one or merely economic like the German one. It was a group expansion as a society and, moreover, it was a very enterprising and vigorous society. "But precisely because Britain was so exemplary a case, it was atypical"[29] although paradigmatic from an imperialist point of view, with all the negative connotations that that entails.

French and Italian imperialism were different as we will see afterwards since, although Great Britain was their model, those countries acted in different contexts, both nationally and internationally. In any case, British colonialism itself would change during the century as imperialism in rivalry with another is different to solitary imperialism.

Wesseling distinguishes two phases in British imperialism: in the first place, a solitary empire around the mid-Victorian period, defined as "imperialism of free trade" by the historians Gallagher and Robinson, and, afterwards, a competitive colonialism at the end of nineteenth century.

At the start, expansion was the logical consequence of a vigorous society and that become evident in several ways: emigration and colonisation, trade and investment, transmission of culture and religion, and maritime settlements. Political imperialism, that is, imperialism understood as a political authority's imposition over unknown people, was another modality but it was not the most important one, although it was the most common around the end of the century. According to Wesseling, that was not the consequence of a change in British home politics but rather the consequence of a change in the international context with the emergence of rivals or emulators. England had its own idea of colonialism. The historian James Froude published his work *Oceana* in 1886, in which he passionately defended a British *Commonwealth*.[30]

As for Somalia, this was the "Cinderella" of the empire, which did not pay too much attention to the country, so the British government ordered as gentle an occupation as possible. Nevertheless, plans had to change when the Dervish revolution, led by Sheik Mohamed Abdullah Hassan (called "Mad Mullah"), broke out. Mad Mullah kept them on tenterhooks for 21 years.[31] Despite everything, from the end of the Dervish revolution to the end of the Second World War, the British administration of the Somaliland protectorate was quite negligent, above all if compared with the Italian administration in the South of the country. The English

informal means, that is, commercially and economically, but if necessary formally as well, that is, politically." WESSELING, *op., cit.*, p. 52.

[29]Vide ibid, p. 55.

[30]Quoted by Wesseling on p. 56.

[31]*Somaliland: Democratisation and its discontents*, cit, The crazy Mullah has gone down in history as the first national Somali leader; however, his acts produced the first division amongst the Somalis: on one side, the "Dervishes", those who took sides with the clan-family that "Mad Mullah" belonged to, the Darod, who had no treaty with the British and, on the other side, the Isaaq clan, the majority clan in the North, who sided with the British with whom they had signed a protectorate treaty.

administration had a minimal interest in the economic development of the region and the "indirect rule" system kept the traditional authority system intact. They used the clan chiefs (*'aqiilo*) as mediators between the government and the people. District courts presided over by local *qadis* applied the *Xeer*.[32]

5.2.2.2 French Imperialism

After 1870, France was a decadent European power from an economic and political point of view. The French had also been humbled by Germany, who had snatched Alsace and Lorraine from them. Germany was then their mortal enemy and this resulted in their alliance with England. The only foreign affair for France was the question with Germany but the country was focused on domestic personal political rivalries, not on foreign affairs. The question was how to restore lost prestige and how to regain their international influence.[33] In this sense, the country was divided into those who supported taking revenge against Germany and those who supported overseas expansion as a way to national recovery. From 1870 to 1914, French foreign affairs revolved around this dilemma: a role on the continent or a role overseas?[34] Although the continental question was considered more important, to tell the truth, from 1870 on, France became the second empire in the world, playing an outstanding role in Africa.

France had occupied Algeria in 1830. This occupation was slow but it would become the cornerstone of the empire.[35] The French government expected Algeria to be not a colony but an overseas part of its territory; this is why it was treated as part of the country from 1881.

France had neither colonial vocation nor ideology; it was the army (in fact, the navy) and the intellectuals (mostly geographers, since the "Philosophy of the Earth" was the science in vogue at the time) who sought to fuel the colonialist mentality, however trade circles were less interested.

The two great architects of French colonialism were Leon Gambetta and Jules Ferry. Gambetta was a convinced colonialist obsessed with revenge against Germany. He thought colonialist expansion would be a solution to social problems and a safety valve for martial instincts. His great power came from his presidency of the Chamber and he had a strong influence over Jules Ferry.[36]

[32] *Somaliland: Democratisation and its discontents*, cit.

[33] Re: French colonialism, see WESSELING, *op., cit.*, p. 29 and following.

[34] Vid. Ibid., p. 30.

[35] Vid. Ibid., p. 31. The occupation of Algeria was curious. France owed the Dey of Algiers a large loan which they had borrowed during the Napoleonic wars. In the opinion of the Dey, they were taking a long time to repay the debt. During a conversation on the subject with the French consulate, the Dey hit the latter on the nose with a flyswatter, thus giving the French a reason to avenge the offence and so began the occupation of Algeria in 1930, which continued in spite of the vicissitudes of different French governments.

[36] WESSELING, *Op. cit.*, p. 35.

Ferry was the architect, prophet and apologist of the new French colonial expansion, supporter of modern colonialism which exports capitals not persons. He distinguished three reasons:

1. Economic reasons: he supported modern colonisation, exportation of capital and goods but not persons.
2. Humanitarian: according to Ferry, "superior races" had the duty to civilise the "inferior races". The racism of the approach deserves no comment.
3. Political: the main reason for Ferry. The Alsace and Lorraine question was an incurable wound but that affair could not sentence the French to international paralysis. No involvement in world affairs would mean descending from first rank to the third or fourth.[37]

Pressured by Gambetta, Ferry led the Third French Republic towards colonisation with the Tunisia expedition in 1881.

As we already know, France had created the Port town of Obock on the Red sea coast in 1862. The country was interested in having a base there with a view to the opening of the Suez Canal. This town would be the seed of the future Djibouti, the small colony that would secure France its presence on the important trade route of the Red Sea. At Djibouti, France established a military base that it would maintain until 1977 when this tiny country got its independence.

5.2.2.3 Italian Imperialism

The only problem with Italian imperialism was just that it was Italian,[38] states Wesseling somewhat rudely in reference to Italian imperialism. The most outstanding historical feature of Italy in the nineteenth century was massive migration owing to two convergent circumstances: a deep economic crisis and super-population. This historic fact was called "the migration of the desperate".

In 1870, the Italian peninsula had reached political unity (with great difficulties and contradictions). It had an imperial capital, Rome, but it lacked an empire. The newly-born State had great expectations at the Berlin Conference in 1878, where European powers divided the Ottoman Empire's spoils. However, Berlin disappointed Italy's aspirations because they didn't get any profit, so Italy felt humiliated and defeated. But the fact that France occupied Tunisia, as we know, in 1881, was even more humiliating.[39] This issue caused a change in Italian foreign

[37]("To reason without acting, without involvement in the affairs of the world . . . means descending from first rank to the third or fourth". A quote by Brunschwig to which Wesseling adds: "The reader will note that the second rank was omitted").

[38]This is the extremely harsh opening sentence of Wesseling's study of Italian imperialism. WESSELING, *op. cit.* p. 289.

[39]We must remember that Tunisia is the cradle of the Carthaginian Empire. The city of Carthage, its nerve centre, was founded in C. VIII B.C. and, from there, its influence spread around the

policy: reconciliation with Austria, its mortal enemy, and the consequent entry into the double Alliance in 1882, which then became the triple alliance: Germany, Austro-Hungary and Italy. The result was the resurgence of Italian imperialist ambitions whose main architect was the passionate nationalist Francesco Crespi. Crespi had started out in politics in 1877 and became prime minister in 1887. His imperialist beliefs are supported by some economic and demographic arguments based on historic reflections and imperialist proclamations.[40] Italy "had to" own colonies because of its historic past and because of its economic needs.

As soon as the Suez Canal was opened to navigation in 1869, the Genovese shipping company, Rubattino, conquered Assab Bay on the Red Sea coast in what is nowadays south Eritrea. On July 5th 1882, the Italian government converted this land into the first official colony of the Italian kingdom.

In 1885, the Italians took Massawa Port, situated on the Eritrean coast too, but in the North. From then on, the efforts of the government focused on linking Assab and Massawa, with the purpose of broadening its influence not only on the coast but in the hinterland as well. All these efforts were rewarded with the creation of Eritrea in 1890.[41] But Italy had pinned all its hopes on Ethiopia. However, the Italian army was torn to pieces by the Ethiopian army at Dogali, becoming the only country to be defeated by an African army.

Only during fascism, between 1936 and 1941, was Italy able to "occupy" Ethiopia, incorporating it into Eritrea and Somali, forming a single political entity called Italian Eastern Africa (Africa Orientale Italiana AOI). The king of Italy assumed the title of emperor, while Haile Selassie, the former Ethiopian emperor, went into exile starting a long campaign to raise public awareness against Italy.[42]

The presence of Italy in the Horn of Africa was tolerated by Great Britain and France just because it was a medium-sized power and posed no risk whatsoever for their respective empires. This presence was favoured by Great Britain most of all since, because of its weakness, Italian presence was preferable to French or German on the way to India after the opening of the Suez Canal.[43]

"Although the Italians suffocated traditional feudal practices, they never worried about training or educating a local elite, so the administration and economy were always organised by Italian settlers or soldiers. The decade of British administration

---

Mediterranean but it was conquered and annihilated by the Romans in C. II B.C. after its defeat in the Punic wars – a fact that precisely marks the beginning of the hegemony of the Roman Empire.

[40]WESSELING, H. L., *op. cit.*, p. 291.

[41]Up until then, the territory of this colony had formed part of various empires and kingdoms of the Horn of Africa region. It was the kingdom of Italy that created the colony of Eritrea and kept it, with its current borders, until the Second World War.

[42]BELLUCCI, S., *Storia delle guerre africane: Dalla fine del colonialismo al neoliberalismo globale* (Roma: Carocci, 2006).

[43]BELLUCCI, S., *Storia delle guerre africane: Dalla fine del colonialismo al neoliberalismo globale* (Roma: Carocci, 2006).

(1941–1951) after the Italian defeat in the Second World War allowed the start of the above-mentioned middle class".[44]

5.2.2.4 The Ethiopian Empire

When Europeans arrived in Africa in the nineteenth century, Ethiopia was a decadent empire. Indeed, it was a puzzle of large and small states and regions, with barely any internal cohesion. Nevertheless, the memory of a glorious past, the unifying power of the Christian Church and the presence of an educated elite made the re-launching of this country possible from the middle of the century. In 1855, Kassa, an ex bandit, crowned himself as emperor with the name of Theodor II. He came into conflict with the English and was beaten by them in 1868; he committed suicide shortly afterwards. Then, after the ensuing fight for power, the leader of the Tigre region was crowned as the new emperor with the name of John the IV, in 1872. During his reign, he had to face Egyptian expansion and, also, an enemy at home: his rival for power, the young king of the Shoa region. John the IV would retain power until 1889. The Italians attempted to spread out over Ethiopia but they were defeated, as we have already said, in Dogali. After the defeat, they started the diplomatic machinery rolling to set John the IV up against his enemy at home.

On October 20th 1887, by secret treaty, the Italians recognised the sovereignty of Melenik, the leader of Shoa. In 1889, John the IV died on the battlefield fighting against the Mahdists and the Ethiopian people recognised the king of Shoa as emperor. He reigned under the name of Menelik the second, the *negus negast* (king of kings). He governed until 1908 and turned Ethiopia into a strong and modern centralised state.[45] He provided his country with a state-run post office, a legal tender, a railway and a press. But, above all, he created an army in order to execute an expansionist policy. Even more important was his diplomatic talent. The great powers (not only England and France, but Russia, Austria, Switzerland, Belgium and Italy as well) showed signs of friendship towards him but, although they were sincere, they were not disinterested. Britain and France tried to use Ethiopia in their complex game of reaching the Nile River. Russia and Italy saw the possibility of also playing a part in Africa, however modest. All of them looked upon Melenik, as the British diplomat, J.R. Rodd put it 'as little more than an exceptionally enlightened savage'. In fact, however, Melenik was an extremely skilled diplomat, a man with near Bismarckian qualities.[46]

Italy failed in its diplomatic attempt to establish a protectorate over Ethiopia and had to resort to the army. In view of Italy's advance, Menelik II declared war against Italy in September 1895. The Ethiopian army crushed the Italian army at the battle of Adua with disastrous consequences for them. In the Treaty of Addis Ababa, on

---

[44]BELLUCCI, *op, cit.*

[45]WESSELING, p. 293.

[46]So WESSELING describes it in the book we have been citing. See p. 294.

October 26th 1896, Italy recognised the sovereignty and independence of Ethiopia. From then on, the Ethiopian expansion began.

In the dispute between France and England for the control of the River Nile, Ethiopia supported France and tried to advance towards the river but their defeat at Amdurman interrupted their advance. In the treaty signed on May 15th 1902 with England, Ethiopia renounced its claims to the lands adjoining the Nile but gained control over the high plateau. Thus, the modern state of independent Ethiopia was born. Ethiopia then started its expansion towards the Red Sea and, as we have seen, an agreement with England in 1897 gave it the possibility to subjugate a part of Somalia, the Ogaden.

Ethiopia is the only African country that has never colonised (except for the Italian occupation). This is largely due to the interest of foreign powers in using this nation as a buffer state among their respective empires.[47] So Ethiopia became the Christian bastion, the only Free State in Africa that could brag about maintaining direct relations with European chancellors and courts.

5.3 Republic of Somalia: The Way to War

5.3.1 The Decolonisation Process: Somalia and Somaliland, Just Two Tips of the Same Star

In 1940, at the beginning of the Second World War, Somaliland was occupied by Italy. However, only 7 months afterwards, the British regained control until the end of the war. Somaliland and other Somali territories were then put under British military administration. In 1949, the British gave Somaliland back its protectorate status, as before the war, thus preparing its possible independence.[48]

On October 2nd 1959, the United Nations passed the *United Nations Trusteeship Agreement*, by means of which agreement Italy assumed the fiduciary administration of its ex-colony,[49] "Administrazione Fiduciaria Italiana in Somalia: AFIS". The main purpose of AFIS was to guarantee a gradual economic and administrative power transfer to the Somali people. In order to control the activity of AFIS, a body was set up under UN observation, the United Nations Advisory Council, to serve as the link between UN and the Italian administration. But this body did not have much

---

[47] BELLUCCI, S., *op. cit.*, p.

[48] International Crisis Group, "*Somaliland: Democratisation and its discontents*" Africa Report no. 66, 28 July 2003.

[49] At the end of the Second World War and beginning of the Cold War, United Nations considered Italy to be one of the colonial powers, alongside England, France and United States; for this reason, Somalia was put under its rule from 1950 until their independence in 1960. BELLUCCI, S., *Storia delle guerre africane. Dalla fine del colonialismo al neoliberalismo globale* (Roma: Carocci, 2006), p. 86 and following.

influence on the process. In 1960, after 10 years of trusteeship, the UN decided that independence should be proclaimed for Somalia.[50]

No date was fixed, however, for the possible independence of Somaliland but, in the mid-1950s, a Congress was held, the National United Front, which was participated in by associations and political parties. The purpose of this Congress was to ask for the release of Haud,[51] under British control, and also to ask for their independence as a member-state of the Commonwealth. In 1956, Great Britain decided to speed up the procedure for the recuperation of the Somaliland self-government. Therefore, the calendar for independence was gradually synchronised with that of Italian-Somalia, planned for 1960.

I think it is important at this point to stress the fact that the State being conceived by Europe, according to LULING, "was, as in most of Africa, an essentially artificial one, 'suspended above' a society which would never have produced it and did not demand it". Moreover, this 'mismatch' between state and society was at the root of future problems.[52]

In April 1960, the political leaders of Somaliland went to Mogadishu to unconditionally accept the constitutional regulations foreseen for the independence of Italian-Somalia. One month later, from London, the British government agreed to guarantee independence for its protectorate on the condition that traditional clan leaders expressed their support for the decision. On May 19th 1960, the Somaliland Council of Elderly gave its consent. Queen Elizabeth II granted independence on the night of June 25th 1960. However, the country was not prepared for its new condition as an independent and sovereign state. It had only a few graduates and two secondary schools. It had no asphalted roads to connect the main towns and it lacked any kind of industrial infrastructure whatsoever.

Five days afterwards, on July 1st, Italian-Somalia obtained independence. At a joint assembly, both states announced their unification to form the Republic of Somalia.

The unification process was hard enough. From an institutional point of view, there were two different administrations. Italy and Great Britain had left two separate administration systems, different education systems and legal systems in which judicial decisions entailed completely different procedures.[53] The police system, the tax system and the currency system were also different, as was the education of their respective elite and their official languages. Furthermore, there was no economic contact between them.[54]

---

[50] GUGLIELMO, M., *Le ragione storiche del conflitto* (Edizionei Altravista, 2008). See Chapter II.

[51] Haud is the region situated to the South of modern-day Somaliland and to the North of Ogaden. This territory currently belongs to Ethiopia.

[52] LULING, V., "Come back Somalia? Questioning a collapse state", *Third World Quarterly* 18, no. 2 (1997): 288 and ff.

[53] GUGLIELMO, M., *Somalia, le ragioni storiche del conflitto*, second chapter.

[54] GUGLIELMO, M., *Somalia, le ragioni storiche del conflitto* (Edizioni Altravista, 2008).

Colonialism had distorted the clan-alliance system with the intervention of the British and the Italian, who had won over some clan chiefs using them as key elements under colonial administration. At this stage, new currency had also appeared and new social classes had emerged that had come from the countryside to the big town to join the administrative mechanism of colonisation.[55]

Mogadishu became the capital of the new State, while Hargeysa, the colonial capital of British Somaliland, was relegated as a provincial capital far from decision-making centres. Somaliland representatives obtained only 33 seats of 123 that formed the new National Assembly. The President and the Premier, as most ministers, came from the South. However, the new Assembly soon recognised its mistakes and drew up a new Unification Act, in January 1961, backdated to the moment of independence. In any case, the merger was put to the test by holding a referendum on the single constitution. By then, the enthusiasm for unification had already vanished in the North. The Somali National League, led by Egal, boycotted the referendum so only a minority voted and more than half of those rejected the constitution.

In December 1961, a group of Northern officers from Hargeisa carried out a failed coup d'état attempting to gain secession from the South. This was not a serious injury to unity since, in 1967, Egal, the Somaliland leader, became Prime Minister.[56]

Apart from the North-South tension (existing at that time and which would lead both parts to armed conflict and secession), to tell the truth, at the moment of unification and independence, these two Somalias paradoxically coincided in their pan-Somali aspirations. The aspiration of joining all the Somali peoples in a single modern state was a preferential topic on the first leader's agenda.[57] According to article 6.4 of the Constitution of the Somali Republic of 1960, "The Somali

---

[55]RUIZ-GIMÉNEZ, I., *Las buenas intenciones*, cit., p. 64.

[56]International Crisis Group, "*Somaliland: Democratisation and its discontents*" Africa Report no. 66, 28 July 2003.

[57]The political leaders I am referring to came from diverse associations originating from the clans and rapidly converting themselves into political parties due to the imminent arrival of independence and a new constitutional and democratic political regime just like the European ones, above all the Italian one, which had little or nothing to do with the historical political tradition of the Somali people. The most important formations at that time were: La Lega dei Giovni Somali, the first party created in 1947, coming from the first Somali political organisation, formed in 1943, under the British administration during the Second World War; El Hizbia Digil Merifle Somali, also known as Partito Costituzionale Independiente Somalo; La Lega della Grande Somalia, formed from a split at the heart of La Lega di Giovni Somali; La Lega Nazionale Somala, the majority party of Somaliland; and, finally, El Partito Liberale dei Giovani Somali. Of all the political line-ups mentioned, **only the Lega dei Giovani Somali and the Lega della Grande Somalia seemed able to accumulate a national consensus base.** Both parties had the support of the most numerous clan-families, the Darod and the Hawiye. What determined the success of the Lega dei Giovni Somali was their capacity to transversally lead diverse clan-families, not making blood a political limitation, and their capacity to also obtain support from amongst the Saab clan-families (that is, the Digil and the Rahaweyn, the farmers) See GUGLIELMO, M., *op. cit,* p. 38.

Republic shall promote, by legal and peaceful means, the union of Somali territories and encourage solidarity among the peoples of the world and, in particular, among African and Islamic peoples".[58] This wish was symbolically expressed in the five-tipped white star chosen as the national flag. Each one would represent one of the five Somali territories.

5.3.2 *Brief Democratic Experience*

From 1st July 1960 to 29th October 1969, the Republic of Somalia experienced a brief and disastrous democracy. There were two Presidents and three Prime Ministers.[59]

Upon independence, this new State *was given* a liberal constitution, shaped in the likeness of the Italian one of 1948 and impregnated with enlightened principles of classical European liberal democracy, without taking into account at all whether these principles would be able to respond to the realities and demands of the new sovereign state. It must be borne in mind that it was drawn up by a wholly Italian technical committee, made up of AFIS workers and two university teachers. It is a copy of the Italian constitution with the exception of matters relating to unicameralism and the Muslim religion professed by all Somali citizens.[60]

At the time, Italy extolled the virtues of the text was but the new constitution and its principles were soon seen to be inadequate and completely alien to Somali socio-cultural reality. Here lies, perhaps, the main reason for the failure of the Somali democratic experience.

From the point of view of external analysts, the young African republic seemed to be going well, especially after the 1964 elections in which the participation rate was very high, thus giving rise to the positive opinion adopted by international observers who believed that the Somali multi-party system was a reality close to consolidation, it even being considered a role-model for democracy in Africa.[61]

---

[58]DEL BOCA, A., *Una sconfitta dell'Intelligenza* (Roma: Laterza, 1993), chap. 1. The text of the Somali Constitution of 1960 is available at http://www.somalilaw.org/Documents/Constitution1960.pdf.

[59]Aden Absullah Osman, senior member of the Lega dei Giovani Somali, was the first President of the Republic. His term lasted until 1 August, 1967. He was succeeded by Shermarke, from the same political group and who would be assassinated during his term, on 15 October, 1969. From 1960 to 1964, Shermarke had also assumed the role of prime minister. He was succeeded by Haggi Hussein until 1967 and, from 1967 to 1969, by Mohamed Haji Ibrahim Egal, the only prime minister from Somaliland.

[60]Re: this first Somali constitution, along with the country's other constitutional experiences, see AXMED ASHKIR BOOTAAN, "La Costituzione somala dal 1990" in *Transplants Innovation and legal tradition in the Horn of Africa: Modelli autoctoni e modelli d'importazione nei sistemi giuridici del Corno d'Africa*, ed. Elisabetta Grande (Torino: L'Harmattan Italiana, 1995), pp. 131–165.

[61]GUGLIELMO, M., *op. cit.* Chap. II.

The truth is that reality was far from that. Nine years after unification and independence, a bloodless coup d'état would end Somali's only parliamentary experience. The reasons for this failure are many and intricate, and they overlap. In the first place, the absence of a tradition or culture involving political parties played a part. The political parties had been formed on the basis of clan-like associations and frequently represented conflicting tribal positions which, in turn, contributed to increasingly serious social fragmentation, the most serious fragmentation of all being that which existed between the North and South, which explains the failed 1961 coup d'état carried out by ex-British-Somali officials. Another reason can be found in the attempt at achieving pan-Somali aspirations, which led to the first armed conflict with Ethiopia in 1964 on account of the border region of Ogaden. It involved a violation of the 1963 Cairo Principle, which stated that new African nations were obliged to respect the borders formed during Colonialism; this violation brought about the isolation of Somalia from the international community. From another perspective, nepotism, defection and political economic corruption contributed, in no small terms, to the breakdown of social structure.[62] To this picture, we should also add the changes in Russian and American political strategies, with the Horn of Africa as a backdrop, from within the bi-polar logic of the Cold War.

It is not strange, then, that the young Somali democracy should end in magnicide. The second president of the republic, Shermarke, was assassinated by one of his bodyguards on 15 October, 1969. A few days later, on 21 October, General Siad Barre's coup d'état put an end to the democratic parliamentary system of Somalia, with the acquiescence of its citizens.

5.3.3 From Dictatorship to Chaos

After the coup, the Constitution, Supreme Court and National Assembly were suspended and political parties were declared illegal. All powers were assumed by a Supreme Revolutionary Council, presided by General Siad Barre. In 1970, on the first anniversary of the regime, "scientific socialism" was adopted as the official ideology, aligning Somalia with the Eastern block and allowing it to come out of the international isolation it had been in since 1964 and to receive military support from Russia. Also, in 1974, the country became part of the Arab League, hoping to find support in the Muslim world for its territorial claims against Ethiopia, bastion of Christianity in the Horn of Africa.[63] The socialist experience seemed revolutionary but the revolutionary objectives translated into an extreme militarisation of the government and into a progressive concentration of power in Barre's hands. On the other hand, the regime focused on modernising the country and society, attempting

---

[62]GUGLIELMO, M. pp. 58–63.

[63]GUGLIELMO, M., *op. cit.*, chapter 2.3.

to make the change from nomadic to sedentary and from tribal Muslim to socialist.[64] With this aim, Barre preferred to leave aside pan-Somalism for the first few years, in the understanding that the country's most serious problem was its cultural and technological lag. However, his political and economic project did not work, due not to the incapacity of the Somalis to assume new political theories, as BELLUCI points out, but to Barre's own centralising politics: "the cult towards the personality impeded any dialogue and, above all, attempted to discourage any political antagonism, thus moving the people further away from politics and political power. The struggle against tribalism so became a form of political repression.[65]"

The results of revolutionary politics contributed greatly to the widening of already open gaps in Somali society. In effect, the replacement of the Arab language, used in an educational sphere in Koranic schools, with Somali transcribed in the Latin alphabet, on the one hand, and the adoption of a new Westernised family code, on the other, seriously upset the religious class.[66] From another perspective, the regime progressively became more and more militarised. The army was conceived as the only source of national unity; multi-partyism was inevitably associated with tribalism.

In 1977, the grave state of the economy led the dictator to focus again on pan-Somalism in order to distract the attention of public opinion. His target, of course, was Ethiopian Ogaden.[67] Meanwhile, in Ethiopia, the 1974 *Dergue* revolution had removed the emperor Haile Selassie from power and the new regime of President Menghistu was ending relations with the United States and aligning itself politically with the Eastern block.

The Somali irredentist movement was already operating in Ogaden. Since 1976, with the support of Somalia, a strictly centralised, lay movement had been emerging, with the help of Ogaden officials from the Somali army itself. Thus, the Western Somali Liberation Front (WSLF) came about.

The war against Ethiopia was a non-negotiable point for the Somali regime; a full break with Russia, who supported the Ethiopians, ensued. Somalia looked for help and military supplies from the Arab world. However, Siad Barre also suffered condemnation by the then Organisation for African Unity for violating the Charter which still today sanctions the principle of inviolability of the colonial borders.

---

[64]BELLUCCI, S., *op. cit,* p. 96.

[65]BELLUCCI, S., p. 96.

[66]Until 1972, Somali was a language of oral tradition. The Barre regime decided it should be transcribed, using the Latin alphabet for the purpose instead of the Arabic, which would have been much closer to Somali religious and cultural tradition. Therefore, the contrariness shown by the religious class in relation to this secularising, lay manoeuvre is not surprising. The adoption of a new Family Code, in 1974, entrusted to a westernised jurisdiction in the cities and to the Xeer and the Sharia in the peripheries, heightened the displeasure of the religious class. In relation to this, see GUGLIELMO, M., *op. cit.*

[67]Re: the Ogaden war, see GUGLIELMO, M., *op. cit.,* third chapter, third, fourth and fifth sub-epigraphs.

In March 1978, after numerous defeats and without adequate military support, the WSLF retreated from Ethiopian territory, producing more than half a million refugees who flooded Somali refugee camps, causing great problems to the stability and economy in the North of the country. Another serious consequence of the defeat was the coup d'état perpetrated the following day by a group of young officials who could have constituted an alternative to the regime in their own country but who had been sacrificed to the Ethiopian front. The repression was severe: they were executed or obliged to return to Ethiopia from where they had just fled. As the majority belonged to the clan-family of the Darod, Barre used that as an excuse for a series of purges against this family in the army and in the party. That is how a certain clannishness began in the regime's politics. It was a first attempt to portray the opposition as a clan in order to declare it illegal. There were repressions amongst the people of the northern centre (Mudug) where more than 2,000 members of the Mejerteen clan died in 1979, due to the wretched conditions when the government poisoned their wells and killed numerous livestock.

After the disasters of the war, opposition to General Barre's regime began to emerge from both inside and outside the country. The following movements began to grow out of the Somali Democratic Action Front (SDAF), which had been formed in Nairobi back in 1976:

In 1979 in Addis Ababa the Somali Salvation Front (SSF) was formed. It would eventually become the Somali Salvation Democratic Front (SSDF) in 1981. This movement, formed mostly by Darods, was set up by the officials who had attempted the failed coup d'état in 1978, which we have just referred to and which had, as we have seen, severe repercussions in the North.

Religious leaders and members of the merchant class, resident in Great Britain and coming mostly from an Isaaq background in Arab countries, gave life to the Somali National Movement (SNM) in London in 1980. With no precise ideology, its agenda was to construct a political system based on Somali consuetudinary law, that is, the Xeer, and to pay more attention to the Islamic values denied by the regime. Its immediate objective was to overthrow Siad Barre's regime and substitute it with a federally structured democratic government.

In 1986, as a final epilogue to the false nationalism of the regime, the Ogaden National Liberation Front (ONLF) appeared, filling the space left vacant by the Western Somali Liberation Front.

It is necessary to highlight the importance of the United Somali Congress (USC), a political and paramilitary organisation, with Hawiye fundaments, founded in Rome in 1989. Its military wing was formed in Ethiopia. Soon, it would split over leadership, which was disputed between General Ali Mahdi and Mohame Farah Aidid; a split which would be a decisive factor in the definitive spiral towards civil war.

There was only one unarmed opposition movement to the government, formed in Modadiscio in 1989 and named the "Manifest Group", which brought together political, economics and intellectual personalities. This group aimed for a government of national unity which could peacefully substitute the Military Council. In June, 1990, Said Barre arrested 45 members of the Group, imprisoning some of its members and executing others. A huge popular outcry obliged the release of some of them.

In 1988, repressions in the northern area were such (the city of Hargisia and others with a majority of Isaaq inhabitants were destroyed) that they were practically in a state of civil war. The SNM was fighting against all the non-Isaaq clans armed by the regime. 300,000 Isaaq fled to Ethiopia as refugees.

With the declaration of peace with Ethiopia that same year, the WSLF was definitively left to its fate. That marked the end of a vicious circle in the civil war, which alongside the collapse of state institutions and the economy destroyed Somalia even before the fall of Said Barre, which happened in January 1991.

The oppressive tendency of the regime, with the use of latest generation armoury, had caused unprecedented violence, death and destruction and culminated in an genuine purging of the clans. On the other hand, Barre had distributed these arms amongst civilians considered to be on his side, increasing the level of militarisation in society, feeding the machinery which would blow up the state apparatus.

January 1991: Said Barre was forced to leave Somalia by the United Somali Congress (USC), led, as we have seen, by the Hawiye clan and the bold and clever General Mohamed Farah Aidid, ex-army official. But inside the two-headed USC, neither of the two heads managed to take charge and stop the violence once started. Civil War broke out. In May of that same year, Somaliland proclaims its own Independence, which has not yet been recognised by any state.

5.4 The Failure of Humanitarian Intervention in Somalia: A Reason for Reflection and Change

Towards the end of 1991, the violence led to famine, making survival difficult for the civil population. NGOs and different humanitarian agencies rapidly set all their resources in motion to sound the alarm to the world. If, initially, when faced with the violence of the conflict, nearly all of them had fled to neighbouring countries to try to continue being of help from further afield, they would soon return to play a paradoxical and more than questionable role. Effectively, humanitarian aid ended up, on the one hand, seriously interfering with economic recovery in Somalia, drowning any chance of local competiveness and, on the other, feeding the war economy, engrossing the material resources of the warlords and militias, thus increasing their capacity for war. Unquestionably good intentions and goodwill were, nonetheless, financing the continuation of the conflict and, what is even more serious, it was the self-same humanitarian aid that was supplying the various armed factions with the means to oppress a population that was starving to death.[68]

[68]See: AUGELLI, E., and MURPHY, C., *Lessons of Somalia for Future Humanitarian Assistance Operations,* Global Governance 1 (1995), p. 345 and following. As the authors point out, 40–80% of the food supplies that entered Somalia at this time is calculated to have been stolen by the various armed groups at war. To this we must add the high rents, taxes and, often, even bribes that the workers and agencies had to pay just to be able to operate in the land.

Furthermore, in order to make westerners aware of the situation, the humanitarian bodies and the media exploited images of the Somali drama, giving rise to what De WAAL correctly called "disaster pornography".[69]

So, given the methods used and their sorry results, one could, in agreement with De WALL, talk of "philanthropic imperialism" rather than humanitarian aid. But the propaganda achieved its goal. Between 1992 and 1995, the most ambitious of the humanitarian interventions led by United Nations that decade, that is, during the so-called euphoric phase favouring coercive humanitarian diplomacy, took place.[70] Yet, taking into account the final balance, it ended up being also their greatest failure.

There were, in reality, two interventions that took place in this country, one after the other, without interruption, both with the authorisation of the Security Council and both in virtue of chapter VII of the San Francisco Charter, claiming humanitarian reasons. For the first time ever, the Security Council considered that a humanitarian disaster in one area of the planet constituted a danger to international peace and security.[71] What happened we know only too well; what would be interesting would be to reflect on the lessons learned and the consequences, not only for Somalia itself but, also, for the rest of the world: the most tragic of all, without a doubt, being the paralysis of the international community faced with the catastrophic loss of life in Rwanda.[72]

---

[69]WAAL, Alex de, "Doing Harm by Doing Good? The International Relief Effort in Somalia", *Current History* 92:574 (1993:May), p. 202.

[70]RUIZ – GIMÉNEZ ARIETA, I., *La historia de la intervención humanitaria.*

[71]Resolution 751 of the Security Council on 24 April 1992, creator of UNOSOM (United Nations Operation Somalia), says: "*Deeply disturbed* by the magnitude of the human suffering caused by the conflict and concerned that the continuation of the situation in Somalia constitutes a threat to international peace and security". The resolution is available at: http://daccess-dds-ny.un.org/doc/RESOLUTION/GEN/NR0/013/32/IMG/NR001332.pdf?OpenElement. Later, the Security Council, in Resolution 794 of 3 December the same year, in which the green light is given to operation "Restore Hope", repeats: "*determining* that the magnitude of the human tragedy caused by the conflict in Somalia, exacerbated still more by the obstacles being created to the distribution of humanitarian assistance, constitutes a threat for international peace and security". Resolution 794 is available at: http://daccess-dds-ny.un.org/doc/UNDOC/GEN/N92/772/14/IMG/N9277214.pdf?OpenElement. Lastly, in the same vein, Resolution 814 of the Security Council which extended the mission of UN in Somalia, giving rise to UNOSOM II, pointed out: "*determining* that the situation in Somalia continues to threaten peace and security in the region". This last Resolution of 26 March 1992, can be found at: http://daccess-dds-ny.un.org/doc/UNDOC/GEN/N93/226/20/IMG/N9322620.pdf?OpenElement.

[72]As Walter Clarke and Jeffrey Herbst have pointed out, "not to admit that the alternative to stronger and better-suited international institutions is the starvation and suffering of millions of people is dishonest. Such future tolerance of disorder was previewed in Rwanda in April 1994 when the world, paralyzed by the Somalia debacle, did nothing as the Hutu government slaughtered upward of half a million Tutsis". In CLARKE, W., and HERBST, J., "Somalia and the Future of Humanitarian Intervention", *Foreign Affairs*, March/April 1996, p. 82. In the same vein, Cherter A. Crocker qualifies Ruanda as "the first victim of the post Somalia backlash", in CROCKER, CH. A., "The Lessons of Somalia. Not Everything Went Wrong", *Foreign Affairs*, May/June 1995, p. 7.

5.4.1 UNOSOM and UNITAF: Operation Restore Hope or United Nations and USA in Somalia

After the fall of Siad Barre in January 1991, both his main rivals, General Aidid and General Mahdi, showed themselves to be incapable of taking power, a highly violent conflict breaking out between them. Faced with an intensifying spiral of violence and difficulties in distributing humanitarian aid, on 23 January 1992, the Security Council in its Resolution 733 asked for an increase in humanitarian assistance from UN and other international humanitarian organisations and decided to apply a general and total embargo on all arms supplies.[73] This is the first of many resolutions taken by the Security Council with reference to the Somali situation which continue in force today.[74]

On 24 April of the same year, the Security Council, through Resolution 751, decided to establish the United Nations Operation Somalia, UNOSOM,[75] and to name the Algerian diplomat, Mohamed Sahnoun, its special representative in finding a peaceful solution to the conflict.

In spite of Sahnoun's efforts[76] and some pacts agreed with the *warlords*, the conflict heightened during this first year, fundamentally due to the hijacking of humanitarian aid and its incorporation into the war economy by the latter, who, given their extraordinarily favourable situation, had a rather weak predisposition to negotiate and restore peace. With this scenario, the Security Council authorised an increase in troops for UNOSOM, from 500 to 3,500. But this decision was taken, as Sahnoun has declared, without consulting the delegation of UNOSOM itself in Mogadiscio, nor the political leaders nor the councils of Somali wise men, as had

[73]Resolution 733 of United Nations Security Council, 23 January 1992, no. 2 and 5, available at: http://daccess-dds-ny.un.org/doc/RESOLUTION/GEN/NR0/013/14/IMG/NR001314.pdf?OpenElement.

[74]An analysis of the successive resolutions of the Security Council on the Somali case, which gradually increased the intensity of the military intervention citing chapter VII of the San Francisco Charter, without request from the country's government, in HUTCHINSON, M. R., "Restoring hope: U.N. Security Council Resolutions for Somalia and an expanded doctrine of humanitarian intervention", *Harvard International Law Journal*, 34 (1993).

[75]Resolution 751 of UN Security Council, 24 April 1992, available at: http://daccess-dds-ny.un.org/doc/RESOLUTION/GEN/NR0/013/32/IMG/NR001332.pdf?OpenElement.

[76]The Algerian Mohamed Sahnoun, who already had a long diplomatic history with the United Nations as well as with the Organisation for African Unity, proved from the outset to be deeply knowledgeable about the situation and about Somali idiosyncrasies, but his preference for a "bottom-up" action plan hit head-on with the "top-down" plans of Secretary General Kofi Annan. Shortly afterwards he was no longer required, in the month of November. Since 1997, he has been Special Representative of the Organisation for a United Africa (now African Union) for the Great Lakes Region. He was co-president (with Gareth Evans) of the International Commission on Intervention and the Sovereignty of States, which wrote the Report on "The Responsibility to Protect" in 2001.

been the procedure up to that point.[77] General Aidid reacted to this by blocking the port of Mogadiscio and holding 500 Pakistani blue hats captive. Sahnoun was forced to resign and was replaced by the Iraqi diplomat, Ismat Kittani.[78]

At that point, USA, who had been staying on the sidelines, decided to intervene. Some analysts attribute this change of policy to the so-called CNN effect, that is, to the impact on public opinion caused by the terrible images of Somali refugee camps in Kenya (the "disaster pornography" mentioned above), and to the media news about Somali attacks on humanitarian workers. According to Ruiz-Giménez, several reasons were combined in the decision to intervene which was taken personally by President Bush (already beaten at the polls by Clinton). Alongside an unquestionable humanitarian drive, other reasons that influenced the decision were the euphoria of the Gulf War and the Operation Provide Comfort, the desire to conclude his term brilliantly and the need to ingratiate himself with the Arab countries.[79] Another determining factor was the American conviction that it would be a quick, easy and low cost operation, in terms of human and political resources. That explains, furthermore, why they would condition their actions to the authorisation of the Security Council, an authorisation which came in Resolution 794/92, where the grounds given for the intervention referred to the collapse of the state, further qualifying it as an "exceptional and immediate response". This Resolution was passed unanimously; the reticence of countries, such as China, India or African states who had previously expressed concern about "setting a precedent which would allow the abuse of humanitarian legitimacy in order to hide the secular interference of powerful countries in the internal affairs of other states", appeared to have mitigated.[80] At that time, the prevalent conviction of the international community was that you could not remain impassive in the face of the human catastrophe in Somalia. Perhaps, the most novel aspect of Resolution 794 is that, for the first time, the Security Council expressly authorised a humanitarian military intervention in a sovereign state without the explicit consent of its authorities,[81]

---

[77]SAHNOUN, M. M., "Prevention in Conflict Resolution: the case of Somalia", *Irish Studies in International Affairs*, 5 (1994). In this paper, the Special Representative of the Secretary General for the resolution of the Somali conflict, recounts the difficulties of coordination and the differences of opinion between the heads of the United Nations. All of this led to an early relief of his position and the loss of many possibilities for peace.

[78]On 8 November 1992, Ismal Kittami, an Iraqi diplomat, substituted Mohamed Sahnoun as Special Representative of the Secretary General in Somalia. Kittami remained in the position until 9 March 1993, on which date he was substituted by the American admiral, Jonathan T. Howe.

[79]RUIZ-GIMENEZ, *op. cit.*, p. 180.

[80]vid ibid, p. 183.

[81]As Kenneth D. Bush pointed out, "The Security Council circumvented the implied requirement to enter a country only with the consent of the government by recognizing the destabilizing potential of widespread famine and continued civil war in Somalia as a threat to international peace. The reticence by some members of the Security Council to intervene in the domestic politics of a state (even a "failed state") without a formal request from political representatives was conveniently overcome through the 'creation' and use of what James Jonah calls a 'legal fiction', a Somali request for UN intervention arrived 'in the form of a letter from the Somali charge d'affaires in

although the majority of member states interpreted the Somali case as a clear exception due to it being a collapsed state.

With this scenario, on 9 December 1992, an international military coalition of 24 countries, led by USA and denominated the Unified Task Force (UNITAF), disembarked in Mogadisco, under the spotlight of the western media. On land, UN-OSOM, the UN mission, and UNITAF, the international coalition, came together. The operation was called *Restore Hope,* and immediately the lack of coordination between one and the other became evident. The United States was in favour of a fast intervention, with the lowest possible cost in political and human terms, and limited to the military protection of humanitarian aid; the UN preferred a more stable and long-term intervention with the aim of advancing the peace process and the restoration of the state. Two models of intervention which were completely different and which, on the ground, were obviously incompatible.

In addition to this friction, the warlords managed to inject the humanitarian aid into their war economy again and to continue their armed attacks outside the coalition's area of influence.

5.4.2 UNOSOM II and the Continuation of the Failure

In March, 1993, a new agreement with the warlords allowed Clinton to announce the evacuation of UNITAF from Somali territory and the transfer of the operation to the UN, for which reason the Security Council, on 26 March, created extended UNOSOM or UNOSOM II in its Resolution 814. It was the first *enforcement* operation to be directly controlled by the Secretary General of the United Nations, besides being the first peace-keeping operation which had been given a mandate to use force in order to achieve the objectives of the UN.[82] The majority of countries backed UNOSOM II in spite of the insistence from some (Spain or Morocco) regarding the exceptional characteristics of the operation given the situation in Somalia; they wished to avoid reminiscences which might connect the operation to the colonial past.

In spite of its cost and scope ($4,000,000,000 and 33,000 soldiers and 3,000 civilians), UNISOM II had serious problems from the outstart which irremediably led to its failure: the lack of logistics and troops for their disproportionate pretension of taking control of the country's entire territory; coordination problems at high level; the negative impact on the Somali economy; but, perhaps, the question which made the execution of the mission most difficult was the anxiety of its components

New York who, in reality, represented no one'". In BUSH, D.K., "When Two Anarchies Meet: International Intervention in Somalia", *The Journal of Conflict Studies*, XVII, no. 1 (Spring, 1997): 8.

[82]BUSH, D.K., "When Two Anarchies Meet: International Intervention in Somalia", *The Journal of Conflict Studies*, XVII, no. 1 (Spring 1997): 9. See also MAKINDA.

regarding their own security more than that of those that they were obliged to protect, namely Somali civilians. Therefore, it is not surprising that, given their behaviour, the Somalis soon began to view them as an occupying force that would interfere in their internal affairs. The situation worsened further with the extreme reaction of Admiral Howe, the new Special Representative of the Secretary General, to the death of 24 Pakistani soldiers in June 1993. In pure Wild West style, the American admiral declared General Aidid "wanted" for the incident, although his guilt had not been established. The deeds also met with a response from the Security Council who, through its Resolution 837 of 6 June, ordered the perpetrators to be captured and put at the disposal of the international courts. The violent response of UNOSOM II, which caused the death of more than 100 civilians, was responded to in turn with new attacks by Aidid's National Somali Alliance.

The question is a very serious one because those who had come to give humanitarian aid, to rebuild the state, to return the country to a state of normality, in short, to bring law and order, ended up not only intensifying the violence of the conflict but also forming part of the conflict itself, just like another faction. From June to October, there were several confrontations with numerous victims on the side of UNISOM, certainly, but there were many more innocent Somali victims who fell in the midst of this madness which, however, was silenced by Western media. On 7 October, President Clinton announced the evacuation of American troops, after 18 of his soldiers died in combat on 3 October. But the evacuation is delayed until March 1994, in an attempt to lessen the harmful impact a hasty retreat would have on his credibility. It is not the failure of UNOSOM or the violations of international humanitarian laws denounced by diverse NGOs which prompted the decision, but Western deaths.

The relief with which the Somalis saw those who they already considered invaders leave is understandable, as is a growing tendency towards Aidid's position. After the evacuation was announced, the latter would proclaim a unilateral ceasefire which allowed the Western troops to retreat peacefully, leaving only UNOSOM troops formed by third world soldiers in Somalia.

The UN remained in the country until March 1995. Shortly before they left, the Security Council suspended the wanted order on Aidid. UNOSOM II became a peace-keeping operation. In March, 1995, UNOSOM II sadly left the country, without resolving the conflict, without stabilising the situation and, of course, without rebuilding the Somali state.[83] Somalia, politically fragmented, was abandoned to its fate and the power of the warlords had been strengthened.[84]

---

[83]To give an idea of the failure of the operation, mainly from a humanitarian point of view, it should be taken into account that the humanitarian costs were much lower than the military costs incurred: 2 billion dollars where the military costs of the USA, with a further 2 billion spent by the UN. Humanitarian costs came to 500 and 800 million respectively. Most of these resources went to paying salaries, home rentals, travel expenses and allowances of international staff. The Restore Hope budget exceeded the total sum corresponding to humanitarian aid received by Somalia since its Independence in 1960.

[84]LULING, V., "Come back Somalia? Questioning a collapsed state", *Third Word Quarterly* 18, no. 2 (1997): 287–302.

5.4.3 Data for Reflection

The resolutions of the Security Council regarding the intervention in Somalia shows an evolution of international rules, more favourable to the legitimacy of humanitarian military intervention, in the sense that intervention would be justified by massive violations of the human rights of affected populations and attacks on humanitarian workers, considering that these instances constitute in themselves a threat to international peace and security, as referred to in chapter VII of the San Francisco Charter. As Ruiz –Giménez has pointed out, the novelty of these resolutions was that "for the first time ever, the use of armed force was authorised for explicitly humanitarian reasons".[85]

On the other hand, the convergence of two different intervention logics was also novel and would lend new meanings and content to humanitarian intervention. On the one hand, an intervention which follows the "expedition" pattern, in which the use of force is limited to the military defence of humanitarian aid; and, on the other hand, an intervention following the "garrison" approach, which intends to rebuild the collapsed states.[86] This latter model is the one defended by the UN, following the new civilising standard which emerged after the Cold War but the majority of states prefer the expedition model. Western countries prefer it because it implies less risk and political commitment; non-Western countries prefer it because it has a lesser impact on their sovereignty and, hence, the risk of being used as a cover-up for neo-colonisation processes is reduced.

Transversally to this convergence of two clashing logics, the two players also disagreed as to the *modus operandi* to be adopted. So, the *up-down* focus of Secretary General Boutros-Ghali would prevail over the *bottom-up* focus, favoured above all by Sahnoun. This up-down focus, as Sahnoun said at the time, would marginalise the peace initiatives coming from within the society and favour the warlords' military logic, which materialised the feared risk that interventions which aim to rebuild a state are masking new colonialist logics which, ultimately seek to impose a given political and economic model as the only valid one, the liberal democracies of the marketplace economy, an imposition which denotes the moral arrogance of the West and removes the rebuilding process from the very society that is meant to be being rebuilt, as it attempts to impose institutions which are alien to its own social and cultural roots.[87]

The most immediate consequence of the failure of the international community in Somalia was the weakening of trust in the efficiency of coercive humanitarian

---

[85]RUIZ-GIMÉNEZ, *Op. Cit,* p. 191.

[86]RAMÓN CHORNET, C., *¿Violencia necesaria? La Intervención Humanitaria en Derecho Internacional,* (Madrid: Ediciones Trotta, 1995), p. 98.

[87]See Christopher BICKERTON "State-building exporting state failure", in *Politics without sovereignty a critic of contemporary international relations*, ed. Chistopher J. BICKERTON, Philip CUNLIFFE, Alexander GOUREVITCH (London/New York: University College London Press, 2007).

diplomacy. The passivity of the UN in the face of the extremely serious problem in Rwanda in 1994 was in the least unrelated to this failure. Later disasters in Bosnia (1995) and Kosovo (1999) also contributed notably to the need to rethink the responsibilities of the UN and the international community with regard to humanitarian intervention. A change of paradigm was becoming necessary. The experience initiated in Somalia contributed to opening the way to two new concepts which reflect the new awareness of humanity that one cannot remain impassive when faced with humanitarian catastrophes whatever their origin, the concepts of "human security"[88] and that of "responsibility to protect". In fact, in 2001, the Canadian government established the International Commission on Intervention and State Sovereignty, co-presided by Gareth Evans and Mohamed Sahnoun, was created. This Commission drafted a very interesting "Report on the Responsibility to Protect".[89] The case of Somalia is precisely one of those which inspired this report, alongside those of Rwanda, Kosovo and Bosnia.

As we pointed out earlier, with the evacuation of UNOSOM II, Somalia would be abandoned to its fate, for the time being . . . and in a worse situation than in 1992, but it would not be the last time the international community intervened in different ways. From that moment on and through initiatives with very diverse origins, the sad list of failed attempts at peace conferences would begin.

To the various political and military factions, the Islamic Courts, and the armed factions (with their varying degrees of Islamism) associated with them, would be added later. The actors in the war are multiplying, power is scattering, the conflict is extending to the other countries of the Horn of Africa who side with one or another. The spiral of violence has increased to the point that Somalia has become the most dangerous country in the world.[90] The conflict has gone on nearly 20 years, so several generations have been affected, and it is increasingly removed from its own origins. But understanding this spiral and the impotence of the Somalis and the international community to stop it will have to be the subject of a separate study.

[88]Regarding the concept of human security, see J. BALLESTEROS, *Repensar la Paz,* Ediciones Universitarias Internacionales, Madrid, 2006. A brief synthesis of the origin and evolution of this term can be found in MORILLAS BASSEDAS, J., "Génesis y evolución de la expresión de la seguridad humana", *Revista CIDOB d'Afers Internacionals*, no. 76: 47–58.

[89]See *The Responsibility to Protect*, Report of the International Commission on Intervention and State Sovereignty, at http://www.iciss.ca/report-en.asp.

[90]This is precisely the title of the article published in Foreign Policy by Jeffrey GETTLEMAN, Spanish edition, no. 32, April/May 2009.

Chapter 6
Human Rights and Changes to the International Legal System. Philosophical Reflections on the (Difficult) Coexistence of International Humanitarian Law and International Human Rights Law

Isabel Trujillo

Abstract Chapter 6 analyses the relationship between International Human Rights Law (IHRL) and International Humanitarian Law (IHL). IHL represents the new version of a part of the old *ius gentium*, in particular its *ius in bello*. Nonetheless, IHL excludes *ius ad bellum*, formerly linked to *ius in bello* in *ius gentium*, in continuity with the Just War Theory. The current IHL is characterized by neutrality, impartiality, and humanity in protecting victims. The practice of IHRL – not the abstract doctrine of natural rights – aims to protect human beings in different contexts and against different forms of vulnerability. In this last sense, IHL could be seen as a part of the former in the case of vulnerability caused by war. IHL and IHRL are often indicated as *lex specialis* (IHL) and *lex generalis* (IHRL). Three criteria are used to test the relations between them: the sphere of temporal application (IHL in war or emergency; IHRL in any situation, but with possible derogation in some cases), the question of the addressee (states and other possible actors in both cases), and the sphere of subjective application (the person protected: victims for IHL; human beings in a certain context for IHRL). At the end of the day, the differences between them are less important than they appear to be and suggest an aspect of incoherence in terms of respect for the human person as a key principle of the international order.

6.1 Introduction. Changes to the International Legal System

International human rights law (IHRL) and international humanitarian law (IHL) can be considered two manifestations of the emergence of respect for the human person as a key principle of the international order. Although the history of these

I. Trujillo (✉)
Philosophy of Law, University of Palermo, Palermo, Italy
e-mail: isabel.trujillo@unipa.it

J. Ballesteros et al. (eds.), *Globalization and Human Rights*, Ius Gentium: Comparative Perspectives on Law and Justice 13, DOI 10.1007/978-94-007-4020-4_6,

two phenomena goes a long way back in time, the essence of both, that is to say the consideration of the person and his or her rights as a good to protect in and on behalf of the international community, can be considered an innovation of the last 40 years.

This datum at any rate coexists with the old model of international law as law of sovereign states, in which individuals are simply "subjects." Even in the 1970s – and it is a prejudice that dies hard in the minds of many lawyers – international law was exclusively seen as the law of states, or, in the last analysis, as intergovernmental law, that is to say the law of whoever commands in the state. It cannot be said that now it is no longer so, at least because – partly rightly – only states are endowed with instruments able to obtain respect for the law and hence to protect human beings.

Another important aspect of the same transformation of international law is represented by the emergence of individual responsibility for crimes against humanity, since the time of Nuremberg Process. The work of others international courts of justice overcomes the initial approach, pushing towards an integration of the international system founded on the rejection of that kind of crimes. They identify a customary justification of that principle, in the framework of a (global) system that includes international and domestic laws. The consolidation of international individual responsibility indicates the trend to a common international law, instead of a law inter-states, and makes clear the possibility that, in addition to states, also individuals can be addressed by international norms and not only subjected to it.[1]

In any case, the fact that at the moment the state is still the undisputed protagonist of international law is also due to another cultural prejudice that it is difficult to wipe out: the idea that law is above all "sanction", i.e. that its essential characteristic is enforcement.[2] In this sense, it must be said that the state is the only entity able to use force legitimately.[3] The fact that the law is also (and above all) a rule able to guide human behaviour is thus moved to the background.[4] In particular – as IHL and IHRL in the context of the legal order show – international law contains rules that look at protecting human beings or goods especially important for human beings (for instance, environment): it is not clear whether, in this case, it is the dimension of enforcement to be a stake, or the aim of empowerment or promotion.[5] All these considerations could have interesting consequences for the way of seeing changes

---

[1]Cf. H. Ascensio, "La responsabilité pénale internationale des individus comme épicentre d'un nouveau jus gentium [Ital. translation. la responsabilità penale internazionale degli individui come epicentro di un nuovo jus gentium]," *Ragion pratica* 22 (2004): 11–22.

[2]N. Bobbio, *Teoria generale del diritto* (Torino: Giappichelli, 1993), pp. 115–145.

[3]M. Weber, "The Profession and Vocation of Politics" (1919), in *Political Writings*, (Cambridge: Cambridge University Press, 1994): 309–393.

[4]The inevitable reference is to H. L. A. Hart, *The Concept of Law* (Oxford: Oxford University Press, 1961) and to the reception of his theory among lawyers.

[5]On this side, it is important to name the 'responsibility to protect.' As well known, it establishes the responsibility (of the state, in first place, but of the international community, in case of 'failed states') to protect people against certain crimes, as genocide (in general, crimes against humanity). See http://doc.operationspaix.net/responsabilite_proteger.pdf.

to international law, but this line of reflection would take us away from the central topic of this paper. The direction taken here implies the idea that international law is able to provide reasons and standards for guiding the actions of states and non-state actors; in particular, IHL and IHRL provide reasons for protecting human beings, without discrimination of race, gender, social class, abilities or citizenship.

The transformation hinted at before is also supported by a discussion and re-examination of the role and scope of the so-called "territorial rights of the state": jurisdiction rights and control of resources and borders.[6] An important aspect of the debate on the topic concerns the role that territorial divisions play in theories of global justice. The push for discussion comes from the need to focus on the role of states in protecting human beings, not only its citizens. For example, there is a constantly growing debate about the less and less obvious difference in treatment of citizens of a country and foreigners, a difference that becomes of great importance if it is considered that states would be the best instruments for protection of rights in the domestic and international context. This kind of considerations is linked to the role of "luck" in distributing citizenship; in others words, to its involuntarily character.[7]

Finally, all these changes are set in a situation in which the weight of international civil society is greater and greater on the world scene. In the framework of international organizations, by civil society we mean the set of groups – not of citizens or governments, families or businesses but NGOs, associations of native peoples, volunteer organizations, foundations – that want to draw attention to fundamental issues for the purpose of influencing the policies and decisions of international organizations. International law suggests standards and parameters to these too, and their perception of the international rules is then crucial.[8]

In this schematic introduction it is necessary to avoid two mirror-fashion defects: that of those who do not perceive these changes, continuing for instance to see in international law only the law of sovereign states; and that of those who exaggerate their scope, for example, banalizing the role of the state in international law or extremely emphasizing the role of human rights in it. It is not plausible that international law can do without the state, but it needs to be observed that its role is the object of re-examinations in the light of global demands.[9] It is not possible to ignore the weight of human rights in international law, but the fragmented nature of international law allows some of its parts to be immune to their influence (at least up till now).

---

[6]Cf. T. Meisels, *Territorial Rights* (Dordrecht: Springer, 2005).

[7]Among others, S. Caney, *Justice Beyond Borders. A Global Political Theory* (Oxford: Oxford University Press, 2005).

[8]It seems that it implies a wider approach to the practice of human rights. Cf. A. K. Sen, "Elements of a Theory of Human Rights," *Philosophy & Public Affairs* 32 (2004): 315–356.

[9]Cf. my *Giustizia globale. Le nuove frontiere dell'eguaglianza* (Bologna: il Mulino, 2007).

The central point of the current reflection is meant to be a comparison between IHRL and IHL in the framework of the international legal order.[10] The two phenomena being examined first of all show us that the law of the international community moves – though laboriously – in different directions from those for which it arose in the modern epoch. At its origin the Law of Peoples within the Roman tradition (so-called *ius gentium*) regulated the relationships between peoples, but also between Roman citizens and others that were not Roman citizens. At the start of the modern epoch, however, precisely when the Law of Peoples was preparing to give rise to international law, the axis moved from the individual to the state; *ius gentium* became Law *between* Peoples (and is therefore better indicated by the expression *ius inter gentes* or law inter states). At that moment, it was the law among the nations that were then coming into being in Europe,[11] though it equipped itself with instruments making it able to include other communities like the American ones that had just been discovered.[12] Today in international law the individual has become important (again), this time regarding his or her wellbeing, his or her survival, though – as has already been said – this tendency does not exclude other competing ones.

The purpose of my paper is to define these two phenomena and to build up some philosophical reflections beginning from joint consideration of them.[13] So first of all the aim is to say what is meant by IHL and by IHRL, trying to perceive their underlying logics or philosophies; secondly there will be an investigation of the differences and points of contact between them. Lastly, there will be a reflection on whether IHL has a greater "grip", greater success, in the community of states and also in international civil society compared to IHRL, and, if the answer is affirmative, on why this is the case.

6.2 The Logic of IHL

The term "humanitarian law" refers to the set of norms aiming to limit war violence and to protect the victims of conflicts. It is the name that is given today to the ancient law of armed conflicts, so-called *ius in bello*, which mainly prohibited the killing of civilians not involved in the conflict. As is well known, law *in* war together with

---

[10]Each of these bodies will be considered as a whole, and not considering the relationship between individual norms belonging to them.

[11]Cf. the item *Diritto delle genti* edited by me, with A. M. Moschetti, in *Enciclopedia filosofica* (Milano: Bompiani, 2006), 3: 2966–2968.

[12]For example Francisco de Vitoria upholds the right of indigenous nations too, cf. F. de Vitoria, *Relectio de Indis* (Madrid: C.S.I.C, 1967).

[13]Among others, the parallel and convergence between the two systems in establishing standards of action for states is illustrated with case law examples in A. Orakhelashvili, "The Interaction between Human Rights and Humanitarian Law: Fragmentation, Conflict, Parallelism, or Convergence?," *The European Journal of International Law* 19, no. 1 (2008): 161–182.

the right to resort to war, or *ius ad bellum*, for a long time constituted the content of the Law of Peoples in the modern epoch (the first form of international law).[14] This content was also present in the ancient and original Law of Peoples together with others, but when states became the protagonists these contents happen to be exclusive.

The famous doctrine of "just war", which partly operated ambiguously as a justification of that kind of law, arose from the need to limit armed action, for example by imposing careful reflection before war was declared (the problem of the causes of the war or *ius ad bellum*), prescribing respect for certain conditions of war management (*ius in bello*), and clearly establishing responsibilities. The requirements of a just war are (since that time, with small changes): just cause, right intention, legitimate authority, final resort, proportional means, and reasonable prospect.[15] It is worth noticing that the medieval tradition of just war aimed to mitigate the harshness of war, to limit specious causes of war. But at bottom was marred by the presupposition that a just war was possible, and perhaps even necessary. It is not a case that this doctrine originated in twelfth-century dialectical thought, and it is the result of the methods of conciliating opposites: some authors wondered how to reconcile the evangelical precept of turning the other cheek with the authorization in Roman law to repel enemies with force.[16] The way of reconciling these conflicting demands was to limit the causes of war and the modalities of its enactment.[17]

The development in humanitarian law closest to us forces us, however, to distinguish it from the ancient law of war as described above. This is first of all because what we refer to today as IHL excludes the right to resort to war. In other words, it does not concern the norms that discipline the possibility of declaring war. This specification is very important to understanding the ambiguity of some uses of the adjective "humanitarian" when reference is made to armed intervention.[18]

[14]It is important not to confuse this notion with the Rawlsian Law of Peoples, that is a derivation or application of his theory of justice to the international context, with some adjustments. Nevertheless, the model proposed by Rawls is very ambiguous precisely in relation to its content: the Law of Peoples seems to reflect, with very small differences, the principles valid in the current legal international system, and then it would seem to reflect the current Law of Peoples in the classic meaning. Cf. J. Rawls, *The Law of Peoples* (Cambridge, MA: Harvard University Press, 1999).

[15]Cf. my "Argomenti vecchi e nuovi sulla guerra. Appunti per un riesame del diritto delle genti," *Rivista Internazionale di Filosofia del Diritto* 1 (2001): 118–130.

[16]Cf. H. J. Berman, *Law and Revolution. The Formation of the Western Legal Tradition* (Cambridge, MA: Harvard University Press, 1985).

[17]About the difference of perspectives between the morality of war and the law of war, see J. Mc Mahan, "The Laws of War," in *The Philosophy of International Law*, ed. S. Besson and J. Tasioulas (Oxford: Oxford University Press, 2010), chap. 18.

[18]Cf. M. Walzer, *Just and Unjust Wars* (New York: Basic Books, 1977), pp. 62 and 108. Cf. also the debate between T. M. Franck, "Humanitarian Intervention", and D. Zolo, "Humanitarian Militarism," in S. Besson, J. Tasioulas (eds.), *The Philosophy of International Law*, chaps. 20 and 21.

The norms of the ancient law of war were directed to the prince that had to decide whether to declare war or not. By contrast, one of the characteristics of IHL is neutrality, that is to say the fact that anyone who makes use of it does not take part in conflicts. This is certainly a remarkable change. If and where there is an ongoing war, IHL authorizes intervention with a series of rules serving to protect victims. The principle of neutrality, together with impartiality, independence and humanity, structurally characterizes those institutions whose aim is to achieve the objectives of IHL, including, for example, the Red Cross and the Red Crescent. A controversial additional problem concerns the intervention of those who try to keep the peace in territories in which there is/was a war. Have they got legitimate titles to use force in peacekeeping operations?

The origins of this law in the form in which we know it today must be sought in the law of The Hague and law of Geneva. As is well known, in 1864 the first Geneva Convention on the protection of the wounded and sick in warfare on land, in particular, was signed up, thanks to the push and interest of Henri Dunant, the founder of Red Cross. Since then different conventions drawn up in Geneva and the Hague have progressively given rise to a body of rules inspired by protection of the person-victim on the basis of the principle of humanity, both in international conflicts and also in internal conflicts.[19] At the moment this body is formed by fourth conventions (1864–1949), some of them renewing old conventions for the protection of wounded and sick in Armed Forces in the Field, at Sea, prisoners, civilians, and two main protocols (1977) regarding the safeguard of victims in international and not international conflicts. In that process, IHL was extended to the protection of the victims of natural calamities, whether caused by man or not. This last development underlies the very logic of IHL: the protection of victims. A further but not marginal problem concerning IHL is that there would be a difficulty in identifying those situations of crisis that are not strictly defined as wars, like riots and arbitrary detentions, or even peacekeeping operations, which equally threaten the survival of people. The best understanding of IHL becomes clear in its role of protecting victims in cases of emergency.

The idea that war is a one of the prerogatives of sovereign states (a territorial right, because usually linked to the protection or enlargement of the territory) – that has characterized the old law of war for a long time – has been eliminated by the end of the Second World War. At that time became clear that the most important goal of the international law regarding wars had become their prevention rather than their regulation. The Charter of United Nations recognizes only one condition in which war could be legally permissible, i.e. in self-defense. It means for our proposal that *ius ad bellum* is strongly limited, even though it does not disappear at all: the UN Charter article 51 speaks of the natural right to defense, but at the

[19] P. Benvenuti, "Riflessione sul carattere universale del diritto internazionale umanitario e del movimento di Croce Rossa e Mezza Luna", in *Assistenza umanitaria e diritto internazionale umanitario*, Proceedings of Seminario Internazionale (Milano 2000, Editoriale Scientifica, Napoli 2001): 40–56.

same time indicates the expiring time of it, when the Security Council will have taken the measures to warrant peace and security in the international community. The same article is ambiguous in the part in which it establishes that the right to defense does not preclude the power of the Security Council to enact actions to maintain or establish international peace and security. The solution of this problem is not determined: among the means, it is not necessary to identify the resort to armed force and the coercive measures.

Even though for some time parallel, these two stories (the IHL development and the UN Charter) do not coincide. In addition to that, in the perspective of the UN Charter, *ius in bello* is in any case needed, given the number of internal and international conflicts that have occurred since its proclamation.[20] From this point of view, it is the consideration of war as a possibility, and of IHL as a remedy, that permits to find a unity in the in international law.

6.3 The Philosophy of IHRL

Regarding human rights, reference will be made here not to the idea (or the philosophical or political doctrines) that every human being has some rights that are not subject to the sovereign's will, but to the practice of this idea as present in our current international legal order.[21] So we will be talking of human rights as a legal phenomenon, distinguishing them – though historically they are connected to them – from natural rights.[22]

In this sense, it is necessary to identify the birth of IHRL in the years after World War Two, with the 1945 San Francisco Charter and the 1948 Universal Declaration of Human Rights (a long time after the first Geneva Convention). From the Universal Declaration there started a movement of development of such rights, which schematically can mainly be linked to three different processes: internationalization, specification and regionalization of rights.[23] The process of internationalization made the rights of individuals the content of obligations under states, through the stipulation of international pacts or treaties, and after that in the form of the *ius cogens* content of general international law. *Ius cogens* indicates

[20]Walzer has argued that *ius ad bellum* and *ius in bello* are logically independent. M. Walzer, *Just and Unjust Wars*, p. 41.

[21]The difficulty for philosophers of dealing with sometimes contradictory practical and concrete entities like the practice of human rights rather than with consistent and abstract theories of rights makes authors like Griffin try to square the circle in ways that are not always convincing. See, for example, J. Griffin, *On Human Rights* (Oxford: Oxford University Press, 2008).

[22]F. Viola, "I diritti umani sono naturali?," in *Natura in etica*, ed. F. Botturi and R. Mordacci, pp. 69–92 (Milano: Vita e Pensiero, 2009) and J. Tasioulas, "What is a Human Right [Ital. Trans. Che cos'è un diritto umano?]," *Ragion pratica* 34 (2010): 109–127.

[23]I freely follow the well-known description by N. Bobbio, *L'età dei diritti* (Torino: Einaudi, 1997).

some peremptory norms and *erga omnes* obligations with absolute rather than reciprocal character, such that violation by one part does not allow the other part to suspend its obligations.

The process of specification gradually determined the content of the rights of human beings, clarifying various dimensions of "status" requiring protection, those that are referred to as "generations of rights:" the first generation or freedom rights, the second or social and economic rights, and the rights of the third and fourth generations or rights regarding common goods like peace and development and the rights linked to scientific and technological progress (for example, protection of genetic patrimony). As appropriate instruments for protection of rights the process of regionalization identified regional legal documents like the European Convention on rights, the Inter-American Charter or the African Charter of rights, with their respective instruments for protection. The most evident characteristic of this evolution is that human rights aim at protection of human beings in all or many of their dimensions, in their diversified states of life, and in the different cultural contexts within which protection is to be enacted. In this sense, IHRL has been defined as pleromatic,[24] meaning the idea of a law looking at the highest standard of protection of human beings. For this reason, IHRL needs to look both at the specificity of the contexts and at the adequate instruments required for such protection.

This ambitious purpose has raised major problems: that of its sometimes contradictory justifications, certainly, but above all that of the inflation of rights, issues that have prompted minimalist versions of human rights.[25] There seem to be too many human rights, often conflicting with one another. Elsewhere I have defended the suggestion that in the logic of human rights this characteristic is not necessarily a defect, at least because it highlights the idea that the human being is much richer than the rights that are supposed to protect her, or – in other words – that the legal instrument is insufficient to protect human beings.[26] But undoubtedly this feature makes the practice of human rights more difficult, above all if a comparison is made with IHL, which concentrates on the idea of the victim. In addition to that, a frequent critique of human rights is grounded on the idea that they risk to become a political tool in hands of governments. This is an important point, because it implies the fear that the politicization of human rights work could damage the neutrality of IHL.[27]

[24] L. Lombardi Vallauri, "Diritti dell'uomo e diritto pleromatico," *Ragion pratica* 18 (2002): 155–168.

[25] On the connection between minimalism of rights and military intervention, cf. M. Ignatieff, *Human Rights as Politics and Idolatry* (Princeton: Princeton University Press, 2001) and the criticisms by T. Mazzarese, "Is The Age of Rights at a Turning Point?," *Finnish Yearbook of International Law* 13 (2004): 107–126.

[26] Cf. my essay "I diritti umani come pratica di giustizia globale," *Diritti umani e diritto internazionale* 1, no. 3 (2007): 545–564.

[27] A. E. Cassimatis, "International Humanitarian Law, International Human Rights Law, and Fragmentation of International Law," *International and Comparative Law Quarterly* 56 (July 2007):

Finally, the philosophy of IHRL only works if we assume that they form a more or less consistent body: rights are indivisible. As is well known, that characteristic, together with universality, became part of the common understanding in international law after the *Vienna Declaration and Programme of Action*, in 1993.

6.4 Differences and Convergences

Let us try, then, to compare these two phenomena in detail, in order to notice differences, but also interconnections and convergences. We can use the three classic criteria for the comparison of legal systems: the sphere of temporal application, the sphere of application in an objective sense (the question of the addressee) and the sphere of subjective application (the person protected).[28] Before beginning our analysis it is important to note that, insofar as they are considered different bodies of law, we are implicitly assuming that they are not integrated in a common legal system, even though they both belong to the same (global) legal framework. This kind of theoretical approach depends on a consolidated overview of the unity or plurality of legal systems in legal theory. Their relationship could be – schematically – inclusion, exclusion, or partial overlapping. The phenomenon of legal pluralism, as it is increasingly developing nowadays both in domestic and in international domain, would offer in the future new concepts and tools of analysis.[29]

As regards the temporal sphere of application, we can say that IHL is the law in wartime, while IHRL is valid in time of peace. This difference is mitigated, however, by the suggestion that IHRL also has to be protected in wartime: intuitively the protection of human rights cannot cease in times of war. In this sense, these two systems can be deemed complementary because they can be applied together.[30]

Here it is necessary to recall the doctrine of *lex specialis*, proposed by the International Court of Justice for the relationship between IHRL and IHL. As is well known, even with ambiguities and changes,[31] the International Court of Justice has considered the relationship between IHRL and IHL as *lex generalis* and *lex*

629. This kind of criticism would imply a study on the relationship between human rights and politics. A good explanation of the complexity of human rights in Cf. A. K. Sen, *Elements of a Theory of Human Rights* in note n. 8.

[28]For a detailed analysis cf. G. Venturini, "Diritto umanitario e diritti dell'uomo: rispettivi ambiti di intervento e punti di confluenza", *Assistenza umanitaria e diritto internazionale umanitario*, pp. 57–74.

[29]Cf. F. Viola, "The Rule of Law in Legal Pluralism," in *Law and Legal Cultures in the 21st Century. Diversity and Unity*, ed. T. Gizbert-Studnicki and J. Stelmach, (Warsawa: Oficyna-Kluwer, 2007): 105–131.

[30]An outcome of the Teheran International Conference on Human Rights, 1968.

[31]As is well known, three decisions are relevant: the Advisory Opinion of 8 of July 1996 on the *Legality of the Threat or Use of Nuclear Weapons*, the Advisory Opinion of 13 July 2004 on *Legal Consequences of the Construction of a Wall in the Occupied Palestinian Territory* and *Case*

specialis, applying the criterion used in the domestic legal system in which *lex specialis* (IHL) prevailed.[32] The problem is that application of the criterion in the domestic context is linked to the existence of antinomies in a "unique" legal system. *Lex specialis derogat generali* is not then a standard for solving conflicts between bodies of law,[33] but between single norms of the same system in a concrete case, when two opposed solutions are offered for the same case. The idea of using the criterion for antinomies then suggests – in contrast to what was presupposed in the beginning of this section – that IHL and IHRL form an integrated system. Certainly, this point of view is controversial – or at least this is what international lawyers emphasize –, insofar as international law is characterized by fragmentation.[34] The point is that unity in domestic law is warranted by values, and this seems to be problematic for the international system.[35] Perhaps the International Court of Justice is inspired by the idea that there are too common values in the international community and that the protection of human beings is one of these. Otherwise, it would not be appropriate to apply the criterion for antinomies to different bodies of law.

There is another possible way for explaining this point. Insisting on the specificity of IHL, the International Court of Justice is pointing at the IHL more determined conditions of application in comparison with IHRL; it is recognizing a sort of an "epistemic advantage" of IHL over IHRL. But the point is that the advantage is only a question of degree.

The fact remains, on the other hand, that there is a widespread conviction about the legitimacy of the limitation or even the suspension of human rights in wartime, as emerges from art. 4 of the Pact on Civil and Political Rights or from art. 15 of the Rome Convention, two of the most important instruments for the protection of human rights. In the last document, for instance, article 15 authorizes clearly the derogation of the rights in the Charter in case of emergency and in particular in case of war. This means that it is possible in theory that IHL and IHRL overlap or clash.[36] Nevertheless, among international lawyers it is commonly established that

---

Concerning Armed Activities on the Territory of the Congo (DRC v. Uganda), Judgment of 19 December 2005.

[32]Against the idea of *lex specialis*: "Symposium: The Relationship between International Humanitarian Law and International Human Rights Law," issue of *Journal of Conflict and Security Law* 14, no. 3 (2009).

[33]C. Drouge, "Elective Affinities? Human Rights and Humanitarian Law," *International Review of the Red Cross* 90, no. 871 (2008): 501–548.

[34]The International Law Commission seems to confirm the idea of separate bodies, see the *Report of the Study Group of the International Law Commission*, International Law Commission, Fragmentation of International Law: Difficulties Arising From the Diversification and Expansion of International Law. UN Doc A/CN.4/L.682, 13 April 2006.

[35]On this topic, cf. M. Iovane, *La tutela dei valori fondamentali nel diritto internazionale* (Napoli: ESI, 2000).

[36]A. E. Cassimatis, "International Humanitarian Law, International Human Rights Law, and Fragmentation of International Law," 623–640.

this possibility does not work in the case of economic and social rights[37]: a part of IHRL is then applicable in time of war. On the other hand, from this point of view, this represents an obstacle in considering the indivisibility of human rights.

However, the provision of limiting (some) human rights in time of public emergency is not a pass key for any derogation. It implies that the limitation must be proved and must be justified as proportional to the situation of emergency. It means that the derogation must be tested case by case. From this point of view, the question is not again a problem of relationship between bodies of law, but a question of concrete application of norms.

By contrast, from the point of view of restrictions linked to the conditions of application, IHL appears to be less "controversial", because it admits no waivers. The point is that the requisites of IHL, with the distinction between civilians and soldiers, necessity and proportionality represent the minimum threshold that states cannot go beyond in situations that justify waiving of human rights. There can be no waiving of waivers.[38] Another time, IHL and IHRL are understood together, and, in particular, the practice of IHL depends on its relationship to IHRL.

In addition to that, the derogatory nature of IHRL has to be tempered also with the consideration that not all human rights can be waived: the prohibition of torture would be valid at every times,[39] although it is not so in some cases, for example, in the case of the right to a fair trial. In order to understand this difference, the theory of rights[40] has proposed a distinction between those formulations of rights that appear to have conclusive or peremptory forçe ("no one shall be subjected to torture") and the most generic formulations of a directive nature, like the right to life, to freedom or to safety, which are meant to be applied within certain limits (public order,[41] for example). Peremptory formulations appear to attest the absolute character of certain rights, where "absolute right" means precisely a right that cannot be waived.[42] The only way to justify torture is to argue over what is meant by torture.

---

[37]E. Mothershaw, "Economic, Social and Cultural Rights in Armed Conflict. International Human Rights Law and International Humanitarian Law," *The International Journal of Human Rights* 12, no. 3 (2008): 449–470.

[38]On the proposal that the relationship between IHRL and IHL take the form of the relationship between general law and special law, see the arguments and the jurisprudence quoted in A. Orakhelashvili, *The Interaction between Human Rights and Humanitarian Law: Fragmentation, Conflict, Parallelism, or Convergence?*, quoted in note 13.

[39]Orakhelashvili, in the article quoted above, pp. 174–175, shows that the prohibition of torture that is found in treatises on human rights has been a guide for the interpretation of IHL, confirming the complementarity of the two systems.

[40]Cf. the interesting examination of this point in J. Finnis, *Natural Law and Natural Rights* (Oxford: Oxford University Press, 1992), chap. VIII.

[41]Semantic examination of the notion of 'public order' is truly complex and elusive. It also appears as a limit to the exercise of constitutionally guaranteed rights, but at the same time it is ambiguously connected to such rights.

[42]Cf. F. Viola, "L'universalità dei diritti umani: un'analisi concettuale", in *Universalismo ed etica pubblica*, ed. F. Botturi and F. Totaro, (Milano: Vita e Pensiero, 2006): 155–187.

This kind of problems is based on the idea that, certainly, from the epistemological point of view, it is easier to identify what must not be done than what must be done.[43] The IHL advantage concerns the delimitation of the context within which a right is valid: not at every time but in time of war (even though, as was said before, this is not completely exact). The crucial point is that wartime is in itself a period in which individuals' rights are in serious danger. It is not enough – in the light of recognising rights and dignity to them – to protect human beings against the extreme evils of war. It is too easy to notice a split between protection of rights and war: if the aim is to protect people, in no way and in no case should there be war. Here it is possible to perceive one of the problems presented by IHL in the framework of international law, the same problem as is raised by the old law of war or the doctrine of just war. Its presence makes international law contradictory and inconsistent or, at least, fragmented into different components that can hardly be harmonised. We lack the courage (or the interest) to banish war from international law and then we try to mitigate its consequences, but it is clear that this implies lowering the level of protection of human beings and their rights, or – at the end of the day – even violating individuals' rights. The point is that the person's protection in wartime is a defence against threats or radical evils that ambiguously we are permitting. It is clear that in war the effects are not proportional to wrongs.[44] However, it must be said that, in any case, both IHL and IHRL must be applied in the light of the criterion of the highest level of protection. Both admit minimal standards, but both are compatible with the highest level of protection of human beings, even though their standards are different from one another. In particular, the highest standard of IHL coincides with the full protection of human rights of victims.

Concerning the objective conditions of application (the addressee), it would seem to be possible to say that the bearers of the obligations contemplated by IHRL are the states, while those of IHL lie with any authorities that can intervene in favour of victims.[45] But is this really the case? Certainly from the point of view of international law this appears correct. Yet it is possible to find elements that do not perfectly fit into this reductive reading. One thinks about the fact that the world of international relationships is populated by a multiplicity of actors and that all these seem to be called on to enact the protection of human rights. This demand for respect for the person pervades the law on commercial operators (so-called *lex mercatoria*), international organizations and NGOs. At least, all these people and

---

[43]It was Plato that first pointed out the epistemological priority of injustice over justice. This approach was developed more recently by J. N. Shklar, *The Faces of Injustice* (New Haven/London: Yale University Press, 1990).

[44]See C. Kutz, "Fearful Symmetry", in *Just and Unjust Warriors*: *The Moral and Legal Status of Soldiers*, ed. D. Rodin and H. Shue (Oxford: Oxford University Press, 2008), chap. 4.

[45]Articles 4 and 96 of the first Additional Protocol to the 1949 Geneva Conventions, 1977.

bodies are susceptible of violating IHRL.[46] We could establish that anyone that can violate IHRL is addressed by IHRL. In a parallel way, for instance, all those that could be involved in warfare are addressed by IHL: conduct and responsibilities of belligerent nations, neutral nations, groups and individual engaged in warfare are defined by IHL.

A possible difference between these two systems is to be traced in the idea that in regionalization IHRL finds an important instrument for their enactment, while IHL has a markedly universal vocation, to such an extent that differentiations of its guarantees in relation to different contexts would be unthinkable.[47] Here, once more a clarification is required, in order to emphasize the underlying logics. The regionalization of IHRL does not mean allowing restriction of protections in certain contexts, but rather reinforcing such protection, with the search for specific instruments for specific contexts. It is not by chance that in IHRL application of regional instruments does not preclude other (international) instruments, but becomes compatible with them.

As for the subjective field of application, the protected subject, it can be said that IHRL defends people against states, while IHL protects victims against their own state, or against the "captor" state (in the case of prisoners), or against the occupying state. Initially, IHL protected the citizens of a belligerent state against the violence of the adversary state. Since 1977, with the Additional Protocols to the Geneva Convention, the protection has indiscriminately concerned foreigners, citizens, refugees and the stateless, so in the context of IHL it is not possible to discriminate in relation to citizenship. This is a strong point of IHL: the simplification that results from the identification of its purpose in the protection of the "victims" or "protected" persons. By "victim" we mean a very specific person, at least in a war situation: anyone who is not a combatant, sick and wounded or someone that is menaced in an emergency. This specification in the definition of the subject to be protected – and its complete independence from citizenship – seems to be the most remarkable advantage of IHL over IHRL, which instead propose to protect the child, the woman, the elderly person, the accused in a trial and so forth, in a context in which states are important in order to protect individual rights. The protection of the person-victim is only "easier" from the epistemological point of view, that is to say in order to identify the subject to be protected. By contrast, regarding rights, it seems to be difficult to understand clearly who the subject to be protected is.

This difficulty derives – I believe – from a certain way of conceiving rights in the light of negative freedom rights, as defined by liberalism or libertarianism: they are supposed to be a model for human rights. From this point of view, the

[46]Cf. recently F. Marrella, "Regolazione internazionale e responsabilità globale delle imprese transnazionali," *Diritti umani e diritto internazionale* 3 (2009): 229–258.

[47]G. Venturini, *Diritto umanitario e diritti dell'uomo: rispettivi ambiti di intervento e punti di confluenza*, quoted in note 28, p. 65.

first generation of rights corresponds to an image of the subject of the rights as a possessive individual and an owner, who wants to be protected in his or her freedom from interferences, to such an extent that rights are defined as "shields."[48] But this explanation has very limited scope. In fact, the same cannot be said of the other generations of rights: the second, the third and the fourth, which require positive interventions. It is not by chance that the individuals to whom the rights are recognized are, tautologically, believed to be individuals "needing" protection[49]: against poverty, hunger, the dominance of the holders of scientific and technological progress, of polluters and manipulators. In the same perspective, IHL can be seen as the right to protection from the arrogance of warlords. If this is true, our arguments can go in two directions: one tending to deny that freedom rights correspond to the idea indicated above[50] and one following the logic of the indivisibility of rights, which considerably complicates that exclusively negative idea of freedom.[51]

The idea that the holders of rights are to be protected does not correspond so much to the image of a shield for an individual that is already strong, but rather to awareness of vulnerability. Obviously this is not a new thing of our own era or of the idea of the human being in our time.[52] In our own time, however, we can notice some specific nuances relating to the *scope* and the *perception* of the vulnerability of human beings. According to this view, it can be maintained that IHL is only a part of IHRL: the case in which vulnerability is determined by war or by other forms of emergency. Obviously, harm caused by war is not the only kind of vulnerability that afflicts human beings, even though it is one of the most serious kinds. From this point of view, there is a (philosophical) continuity between IHRL and IHL.

In conclusion, the obligations relating to IHRL, which proposes to protect the human being in his or her manifold dimensions, appear to be difficult to satisfy, but they are no less important. From the legal point of view, concerning the difficulties in identifying its beneficiaries, IHL apparently proves to be less problematic than IHRL. But if its starting point is possible war, the recognition of IHL confirms a sort of schizophrenia in international law: IHL for the protection of the victims is necessary because international law itself allows war, though containing specific norms for the protection of human rights which are incompatible with a state

---

[48]As is well known, the idea of rights as a shield is found in H. L. A. Hart, *Essays on Bentham. Studies in Jurisprudence and Political Theory* (Oxford: Clarendon Press, 1982).

[49]V. Held, "Rights", in *A Companion to Feminist Philosophy*, ed. A. M. Jaggar and I. M. Young, (Oxford: Blackwell, 1998): 500–503.

[50]This is the belief expressed, for instance, by S. Holmes and C. R. Sunstein, *The Cost of Rights. Why Liberty Depends on Taxes* (New York: Norton, 1999).

[51]A. K. Sen, *Resources, Values and Development* (Cambridge, MA: Harvard University Press, 1984).

[52]The topic of vulnerability is dealt with in a deep and fascinating way on the subject of a particular age in western history by M. C. Nussbaum, *The Fragility of Goodness. Luck and Ethics in Greek Tragedy and Philosophy* (Cambridge: Cambridge University Press, 1986).

of war. We come back again to the typical problems of the doctrine of just war. The challenge – from the legal point of view too – would be to equip international law with instruments able to resolve the controversies peacefully and with instruments to guarantee human beings protection in the highest level, and doing it the aim of IHL will be achieved anyway. It will be still necessary to use IHL against different threats that we cannot in any way avoid, such as proper natural calamities.

Part III
European Union as a Referent for Peace and Development

Chapter 7
Europe's Path to Public Reason

Francesco Viola

Abstract Chapter 7 highlights how addressing public issues publicly is a main target of European institutions, considering their commitment to the identification of shared values and the protection of rights. In consideration of this, it is reasonable to ask whether the "Public Reason" set forth by Rawls can be somehow applied to Europe's current perspective, understanding it to be the ruling criterion governing public issues. A major obstacle is to be found in the anti-pluralistic attitude which is widespread across the European states. However, constitutionalism, which is nowadays widely rooted on a global scale, makes contemporary political communities to characterize by disagreement and by the need of a new order of liberties, which is essential for the existence of any political community. In such a context, Rawlsian "Public Reason" does not show up as a useful tool, while Habermas' "Discourse Ethics" seems to be more promising, considering its openness to learning evolutionary processes. Nonetheless, this appeal to public reason needs to maintain its practical character, and should not lead to a philosophical or theoretical dispute, in order not to revive the anti-pluralistic and intolerant attitudes of the European cultural tradition. A European public consciousness should stick to two conditions: respecting the varieties of national cultures and distancing itself from the forms of the national state.

Europe is witnessing a critical moment in terms of its future. It is a moment when it is shifting from a merely economic and legal integration to a cultural and value integration. It was believed that politics was to render this in-depth transformation of the European Union possible. The constitution is a merely political action.

F. Viola (✉)
Philosophy of Law, University of Palermo, Palermo, Italy
e-mail: Francesco.Viola@unipa.it

J. Ballesteros et al. (eds.), *Globalization and Human Rights*, Ius Gentium: Comparative Perspectives on Law and Justice 13, DOI 10.1007/978-94-007-4020-4_7,

But the difficulties which the European constitution is encountering lead us to question whether we truly need to solve the political matter beforehand to reach the integration of values. Isn't the contrary true?

Most likely, all the available channels should be simultaneously activated since Europe is a collective enterprise of individuals, peoples and governments; it is a historical scenario being formed in all of its crucial aspects. That's why (also for ignoble personal interest) I do not agree with Habermas when he claims that the future of Europe depends more on the specialised debates of economists, sociologists and political commentators than on the juridical debates of legal philosophers.[1] It's a matter which leans on all energies and cultural resources, to the exclusion of none. After all, if we look at the preamble of the institutive Treatise of the European Community (also covered and expanded in the preamble of the Treatise of the European Union), we can easily state that the founding fathers did not have an exclusively economicist notion and substantially aimed towards much more.[2]

In Europe, that is committed to defining shared values and guaranteeing rights, publicly addressing issues is even more necessary to ensure that common decisions are reached. Due to this, it is reasonable to ask whether the "Public Reason" set forth by Rawls can be somehow applied to Europe's current perspective[3] given the fact that it is the ruling criteria governing public issues.

It should be remembered that the assumption behind Rawls' Public Reason is the pluralism of "comprehensive doctrines"[4] and it has as an objective the common notion of political justice in a pluralistic regime.

It may seem easy to reach a solution on European Public Reason, since undoubtedly this idea has its roots in the origins of modern thought and dates back at least to the breakdown of the Western religious unity and religious wars, that is, during the first significant form of pluralism of comprehensive doctrines which appeared in the Western world and chiefly in Western Europe. However, the link between the anti-pluralism of the Hobbesian state and the Westphalian principle of the mere coexistence of states, has prevented European political culture from facing the pluralism argument in a pluralistic fashion.

---

[1]Cf. J. Habermas, "Why Europe Needs a Constitution", *New Left Review* 11 (Sep–Oct 2001): 5–26.

[2]Cf., for example the views by J. H. H. Weiler, "The Autonomy of the Community Legal Order – Through the Looking Glass", *Harvard International Law Journal* 37 (1996): 435–436.

[3]J. Rawls, "The Idea of Public Reason Revisited", *University of Chicago Law Review* 64 (1997): 765–807.

[4]A doctrine "is comprehensive if it contains notions on human values, on the ideal personality, on the ideal family relations and associations and on many other things which should lead our conduct and actually our entire life; it is entirely comprehensive, in fact, if it covers all the recognized values and virtues under one system rather precisely structured; it is only partially comprehensive when it comprises various non political values and virtues, but not all, and is structured in a less solid way" (J. Rawls, *Political Liberalism* (New York: Columbia University Press, 1993)). We can hence deduce that a comprehensive doctrine tends to let the justice of institutions depend on values which are not strictly political and even on more global world visions.

7.1 The Anti-pluralism of European States

Europe is witnessing a critical moment in terms of its future. It is a moment when it is shifting from a merely economic and legal integration to a cultural and value integration. It was believed that politics was to render this in-depth transformation of the European Union possible. The constitution is a merely political action. But the difficulties which the European constitution are encountering lead us to question whether we truly need to solve the political matter beforehand to reach the integration of values. Isn't the contrary true?

Most likely, all the available channels should be simultaneously activated since Europe is a collective enterprise of individuals, peoples and governments; it is a historical scenario being formed in all of its crucial aspects. That's why (also for ignoble personal interest) I do not agree with Habermas when he claims that the future of Europe depends more on the specialised debates of economists, sociologists and political commentators than on the juridical debates of law philosophers. It's a matter which leans on all energies and cultural resources, to the exclusion of none. After all, if we look at the preamble of the institutive Treatise of the European Community (also covered and expanded in the preamble of the Treatise of the European Union), we can easily state that the founding fathers did not have an exclusively economicist notion and substantially aimed towards much more.

In Europe, that is committed to defining shared values and guaranteeing rights, publicly addressing issues is even more necessary to ensure that common decisions are reached. Due to this, it is reasonable to ask whether the "Public Reason" set forth by Rawls can be somehow applied to Europe's current perspective given the fact that it is the ruling criteria governing public issues.

It should be remembered that the assumption behind Rawls' Public Reason is the pluralism of "comprehensive doctrines" and it has as an objective the common notion of political justice in a pluralistic regime.

It may seem easy to reach a solution on European Public Reason, since undoubtedly this idea has its roots in the origins of modern thought and dates back at least to the breakdown of the Western religious unity and religious wars, that is, during the first significant form of pluralism of comprehensive doctrines which appeared in the Western world and chiefly in Western Europe. However, the link between the anti-pluralism of the Hobbesian state and the Westphalian principle of the mere coexistence of states, has prevented European political culture from facing the pluralism argument in a pluralistic fashion.

The European modern state is not born pluralistic, but sees in pluralism a type of particularism which most menaces its unity and sovereignty, swallowing it up into medieval fragmentation. The focus of political power is not pluralistic by definition, nor is the process of differentiating temporal from spiritual power, whereby political power has become independent from ecclesiastic power.[5]

---

[5]However, the entire roots of this differentiation process are medieval. Cfr. H. J. Berman, *Law and Revolution. The Formation of the Western Legal Tradition* (Cambridge, MA: Harvard University

The void left by the expulsion of ecclesiastic power has been filled by the consolidation of common values shaped under the watchful eyes of political power, but not always have they come directly as a product of its own will. Thus, slowly, the European states have become nation-states where the pre-existing obligations of a pre-political nature, rooted in civil society, have been reinforced, modified and absorbed by legal procedures and political decisions.[6] The nation-states are far from being pluralist, something which can easily be established from observing the difficult life ethnic, cultural and ideological minorities have witnessed within them. However, the hardships of these minority groups have been widely compensated by the value legitimation supplied by the nation where the political community found its identity and its own common good, meaning its "completeness".[7] Hence the state was able to supply the justifications to its preconditions without recurring to external ideological powers and without necessarily becoming the direct supplier of values.[8]

If the nation-state is tied to a form of life determined and animated by shared values, the European Community in turn was not conceived as a political community, but as a juridical space common to nation-states which relinquished a small part of their sovereignty[9] in order to cooperate within specific economic sectors. This means that the European Community has not lived its own life, but has survived with the support of the other Member States and their government and jurisdictional bodies. Even though the community laws prevail over national laws, even though the decisions by the Court of Justice are followed by national judges, from a political point of view, the European Community has been parasitical to the Member States. Hence, under these conditions it would be out of question to conceive a European Public Reason.

Finally, while Rawlsian social issues related to Public Reason, tied as it is to pluralism, cannot be applied to those other Member States understood as nation-states, these are even further out of place for the European community due to its lack of political dimension. It can only be stated that in the European juridical arena, consolidated public life profiles of the Member States are projected in a muddled manner and that, moreover, they exert their various degrees of influence depending on their extremely different economic and political weight.

Press, 1983) and E.-W. Böckenförde, *Die Entstehung des Staates als Vorgang der Säkularisation*, in Id., *Recht, Staat, Freiheit* (Frankfurt a.M.: Suhrkamp, 1991).

[6]The usual distinction between two concepts of the nation holds no importance in this instance: one more voluntaristic and artificial (*ius soli*) and the other more ethnic and community oriented (*ius sanguinis*). In any case, the nation represents a common destiny, voluntary or involuntary as may be.

[7]On the nature of completeness of the political community cf. my article "La crisi della politica come comunità di vita", *Dialoghi* 1, no. 1 (2001): 40–49.

[8]We are well aware of what occurred when the nation-state exercised its political powers under ideological form. The totalitarian states of the twentieth-century represent the extreme and violent exacerbation of the anti-pluralism already found in the nation-state.

[9]The community represents, as sustained by the Court of Justice in the Van Gend & Loos case, "a new legal order of international law for the benefit of which the states have limited their sovereign rights, *albeit within limited fields . . . .*". The italics are mine.

These observations explain the widespread trend, especially by philosophers of law and politics, to look at European cultural problems through the lenses of the internal cultural situation, both when describing them and when proposing a way to tackle them. The lack of political consistency in the Community favours the instrumental use of the usual argument which invokes the conformity of European parameters to criticize what occurs in one's own country. But this topic is purely ideological, since there are no European cultural parameters which differ from the mere summation of the lifestyles typical of the single states. In fact, Europe itself is, as it has wished to be, representative of the pluralism of national paths which, for better or worse, have developed on their own and which still tend to evolve based on their own internal resources. By this, we do not wish to undermine that cultural tie represented by the common European philosophical tradition, but this becomes increasingly less significant the more the member states increase with their diversity and, in any case, – as mentioned before – it is inappropriate to generate an authentic Public Reason and related issues with the aforesaid conditions.

7.2 The Evolution of Contemporary Constitutionalism

Is the situation destined to change with the onset of European citizenship? Actually Public Reason is a reason related to citizens and people rather than to states. However, European citizenship – as we know – implies the national one. This means that Europe's journey towards Public Reason shall have, necessarily, to go by way of the states and cannot be separated from them. It needs to be ascertained whether these states are open for an internal transformation which does not repudiate their past, but which knows how to implement the new ideas on law and politics. I believe that the future of Europe still depends on the evolution of a modern state and on its capacity to survive. Economy has done enough, but now it is up to the political, legal and social awareness.

We will now attempt to provide some evidence which – in our opinion – can help the European Union towards a shift of direction and if properly exploited, can configure a uniquely *European* route towards Public Reason.

The first, due to its importance, concerns the evolution of contemporary constitutionalism. To this purpose I must be quite sketchy especially considering the complexity of the subject.

Within the framework of the national state, the constitution has represented the program of a community life, a political formula where the fundamental directive values have been outlined together with the institutions which should implement them in social practice.[10] Consequently, constitutionalism has received

[10]For an overview of the evolution of constitutionalism cfr. M. Fioravanti, "Costituzione e politica: bilancio di fine secolo", in *La nuova età delle costituzioni. Da una concezione nazionale della democrazia a una prospettiva europea e internazionale*, ed. L. Ornaghi, pp. 49–67 (Bologna: Il Mulino, 2000).

a state-centred interpretation and has been dominated by the image of institutions which are at the centre of political, social and economic life and which grant univocity and stability to the interpretation and application of the essential constitutional values. The current concepts of the people, popular sovereignty, citizenship, equality, recognition and democracy tend to imply that the nation-state complies with the unitary, legal, political and centralised system.

This type of constitutionalism was unsuited to face the challenges of pluralism coming from inside and out. It is not only a problem of keeping in mind the increasingly growing number of immigrants, exiled and refugees, but also the growing disagreement between citizens on the way of interpreting and putting into practice the constitutional values. The constitutional state should now move away from the nation to face three types of conflict: interests related to control and distribution of resources, cultural identity and values.[11]

When the state no longer identifies with the nation, the way the constitution is considered starts to change: no longer is it a program of social life which inspires the state institutions, but it becomes the language of disagreement.[12] Disagreement becomes legitimised, the opposite positions are corroborated, yet at the same time the public debate is administered based on practical reasoning and not on mere imperious acts. The constitution (now even in countries with well-rooted state traditions) acquires a supremacy towards the state itself purely and simply reduced to an institutional and procedural device which grants legal forms to the constitutional discourse. Constitutional law governs over and above the exercise of government powers. "Constitutional law and discourse is no mere reflection of a prior political order or process, but is recursively implicated in the elaboration of that order."[13]

The effects of this univocal transformation and exclusiveness of the legal system are important. Externally, while the nation-state posed itself as a separate and incommunicable entity, the supremacy of constitution over the state now allows for commonality and dialogue between the current constitutional values in some way also present in other constitutions (especially those related to human rights) and hence leaning towards universality. Seen from within, the constitutional matter should shape the intra-state claims and the sub-state movements regarding relations between the different groups (national, ethnic, territorial, religious, of gender, language or other differences) so that these are not segregated, separating them from the common matter, but aiming to include and legitimise the majority of visions and interpretations of the same constitution through a mutual recognition and competing

[11]C. Offe, "'Homogeneity' and Constitutional Democracy: Coping with Identity Conflicts through Group Rights", *Journal of Political Philosophy* 6 (1998): 119–124.

[12]Cf. J. Waldron, *Law and Disagreement* (Oxford: Oxford University Press, 1999).

[13]N. Walker, *The Idea of Constitutional Pluralism* (Fiesole: European University Institute, 2002), p. 33.

process of negotiation.[14] This means that we have to abandon the assumption that cultures worthy of recognition must already have a "national" dimension or be accomplished and autonomous forms of life, and also that we have to challenge the idea, accredited by an ideological use of the principle of self-determination of peoples, that nations must for this very reason be recognized as states.[15]

In a pluralistic order, the legal system is no longer presented as a uniform process of law by a sole centre of authority, but as the result of an unstable interrelation between multiple types of authority or claims of authority set in different areas or in different processes inside or outside the state itself.[16]

In other words, constitutionalism is not limited to fostering the law-state on the value level, but forces it to be open to multiple solutions or settlements which govern any disagreement, transferring it from rules to legal practice. Law should administer and curtail disagreement which has, equally, been legitimised by the law itself.

7.3 The Order of Liberties

In terms of the nation-state, contemporary constitutionalism introduces – as already mentioned – an appeal for universality which counterposes the national idea intended to specify and identify a human group in distinctive and separate lifestyles from the rest of humanity. "We" are separate from "They", and sometimes "We" are opposed to "They". However, the objective of the European Community does not aim to cancel the distinction between "us" and "them", but to avoid it degenerating into a conflict. Is a community made up of "us" who are, at the same time, different one from the other and yet in dialogue and working together, possible? The introduction of the universalist approach seems to weaken the possibility of the distinction itself and to break up the concept of "us". That's why it short circuits the national dimension and halts European unity. This is a crucial point and the future of Europe depends on it, but also that of the national community policies.

Constitution, with the achievement of the personalist principle, becomes the foundation for the claim of special identities, where people find their own identification and self-esteem. However these people request such an identification from

---

[14]The diagonal set-up of identities and affiliations should also be added to the aforesaid. This is well evidenced by A. Sen, *Identity and Violence. The Illusion of Destiny* (New York: W.W. Norton & Company, 2006). Consequently the Rawls concept of "a totally comprehensive doctrine" becomes outdated since each individual must respond to appeals deriving from different allegiances.

[15]*Contra* cf. Ch. Taylor, *Why Do Nations Have to Become States?*, in *Reconciling the Solitudes: Essays on Canadian Federalism and Nationalism*, ed. G. Laforest (Montreal/Kingston: McGill-Queen's University Press, 1994). More generally cfr. J. Tully, *Strange Multiplicity. Constitutionalism in an Age of Diversity* (Cambridge: Cambridge University Press, 1995), p. 9 ss., where other enlightening critical observations on state-centred constitutionalism and on the need for transformation can be found.

[16]N. Walker, *The Idea of Constitutional Pluralism*, p. 30.

the public *ethos* which is the heritage of the nation. What is different, in order that it should be recognised as such, should already be a part of what is common. The recognition of particularity is only possible on the basis of an overall sense of significance.[17] The public *ethos,* in turn, is not a universal horizon, since the political community where identification is required, is an expression of a collective identity which is also particular. Actually, from a certain point of view, public *ethos* appears to be much more particular than the same identity requests which, coming from the consciousness of people as such, adds up to more or less universal tension.

For instance, all homosexual couples ask to acquire the same rights as heterosexual couples in all instances. Here, we are talking about a universal public recognition of sexual identity with a constitutional foundation. However, this recognition is requested by a political community where the public *ethos* (itself constitutionalised) is traditionally based on the heterosexual family.[18] The more intense the strength of public *ethos*, the more identities feel threatened by standardization. However, if the political community is less cohesive, the source of recognition weakens and the identities do not receive the confirmation they sorely need. Consequently, identity requests, whether they be particularistic or universalistic, grow out of proportion, becoming increasingly more uncertain and conflictual between themselves. Identities wish for a political community which is both weak and strong, weak in terms of reference values and strong in terms of recognition capacity. In fact, the progressive crumbling of common ethics has not only weakened the political community, but has also rendered recognition barely significant.[19] Quite literally, if the main social network ceases to exist, its extension to different relations between couples loses significance.

This is only one example of the conflict between universalistic identity claims applicable within consolidated community frameworks which are simultaneously asked to survive and deny themselves. Reference to the family, however, is especially indicative of the more general support value which is fundamental for a political community deserving of the name. The function of this value is to put order within various individual freedoms which, if left to themselves; would swarm into a multitude of uncontrolled demands, certainly exciting for the anti-prohibitionists, but substantially destructive for the social-political community. Freedom which is not tidily ordered does not allow for social life to be understood and becomes just a mere *modus vivendi*.

---

[17]Once their identity has been recognized, the subjects learn something more about their own special identity and should abandon the stage of ethicality already attained, to reach recognition of a more important pattern of their identity. Cfr. A. Honneth, *Kampf um Anerkennung. Zur moralischen Grammatik sozialer Konflikte*, (Frankfurt a.M: Suhrkamp, 1992).

[18]Cfr., for instance, J. Coughlan, "La reazione delle Chiese alle proposte legislative di riconoscimento delle unioni omosessuali in alcuni Stati dell'Unione Europea", *Daimon*, no. 4 (2004): 279–307.

[19]After all recognition of the broadest identities threatens the same existence of a common public *ethos*. This, in fact is the Honneth theory: Hegel did not reach the objective of determining an abstract horizon of values open to various life intentions without losing the solidarity strength behind the formation of a collective identity.

It is significant to note that Rawls explicitly refuses to consider liberty the core of his political ideas. The overriding political value is not liberty as such, but from the start, specific forms of liberty (*basic liberties*).[20] For Rawls, liberties are already qualified from their outset and belong to a system or to a family which has to count on its internal order. The further problem of the equality of clusters of liberties attributed to each individual should be added to this.

If we observe how Rawls faces this problem of determining this list of fundamental liberties, especially after the criticism he received from Hart,[21] we should note that the parties in their original position do not decide – as we might expect – first on the contents of this list and then on the principles of justice which govern distribution, but simultaneously on one and the other.[22] This is a confirmation that political values come to light already modelled by law, meaning by rules which outline the extent of their application and their balance, at least in principle.

Actually, when one sustains that a value is "political" this means something else compared to its pure and simple ethical dimension. It means that it concerns the combined life of people which, could precisely be guided by different ideas of what's right. Politics is not simply the place where these personal visions should come together in some way, but a place where an agreement on shared values which structure common life should be reached. That's why Rawls very appropriately believes that the essential political value is justice, meaning what directly governs how law should be implemented. The political theory is the result of the interrelation between ethics and law, which after all represents its main role, that is to say, determining what is right, what must be attributed or granted to those who not only participate in daily life but also to the measure they are take part in it.[23]

Contemporary constitutionalism, then, often generates the question of Public Reason in dramatic terms, since, on the one side, it corroborates individual and collective identity needs intended as in themselves good and not simply as unpredictable preferences and, on the other, thus questions the order of the freedoms so that it is not sufficient that these are considered as good *in themselves*, but it is necessary that they become part of the subject of the common good, meaning goods *for all*. In the public arena, any request for recognition which leads to legislative interventions should be based on the good of all, because only this way can the restriction of liberty which any law entails be justified. Within the logic of *identity*,

---

[20] J. Rawls, *Political Liberalism*, pp. 289–290. Cfr., also, R. Alexy, "John Rawls' Theorie der Grundfreiheiten", in *Zur Idee des politischen Liberalism. John Rawls in der Diskussion*, ed. Philosophische Gesellschaft Bad Homburg and W. Hinsch, pp. 263–303 (Frankfurt/M: Suhrkamp, 1997).

[21] Hart had observed, amongst other things, that the reasons why the parties in their original position adopt precisely those fundamental liberties and not others and set the priorities for these had not been adequately explained. Cf. H. L. A. Hart, "Rawls on Liberty and Its Priority", in Id., *Essays in Jurisprudence and Philosophy*, pp. 223–247 (Oxford: Clarendon Press, 1983).

[22] This fact has been especially stressed by R. Alexy.

[23] Generally speaking, cf. my article *Rawls e il* Rule of Law, in "Quaderni della Rivista internazionale di filosofia del diritto", n. 4, ed. A. Punzi, Milano, Giuffrè, 2004, pp. 179–210.

legislation is often invoked for its symbolic role, both for emancipation purposes or with an aim to compensate for atavic and unjust social exclusions; but the fact still remains that what is liberating for some often implies sacrifices for others, and not only from an economic perspective, but also – more importantly – based on the concept of the good life and common welfare. That's why identity requests should be justified in terms of common good, meaning everything that associated members should accept given unbiased social cooperation as a foundation for common life. This is the role of Public Reason.

7.4 Public Reason and Public Use of Reason

The situation proper to contemporary constitutionalism generally emerges differently depending on the various historical-political frameworks. The high degree of unrest in Europe is mainly due to the consolidation of national lifestyles tied to a long historical process where the religious and civil aspects are woven inextricably. The threat for these, represented by the demands for liberty uncaring of the common good, overturns the Rawlsian views on Public Reason. In fact, while Rawls draws inspiration from pluralism and asks how individuals or groups tied to different concepts of life can build a joint political society, the question in Europe, on the contrary, is posed as to how pre-existing political societies can implement pluralism without being overturned by a progressive and inevitable fragmentation of the public sphere. Consequently, Public Reason should take a different path.

While Rawlsian Public Reason proceeds by way of restrictions,[24] the European one shall have to develop by progressive expansion and inclusion. While Rawlsian Public Reason uses the method of *avoidance* to by-pass the true matters of comprehensive doctrines competing amongst themselves,[25] European Public Reason implies that reasoning is part of rationality and hence cannot totally evade the problem of the truth behind beliefs and demands set forth in the public sphere. It is for this reason that Habermas prefers to cite Kant when speaking of "public use of reason" rather than of "Public Reason". The public use of reason belongs to all rational beings without any restrictions dictated by some form of authority or by other procedures. It represents the use of reason in its entirety. Consequently, the concept of the public sphere is broader than the Rawlsian one, but its dynamic nature is more conflictual than competitive. However, it should be noted that, while for Rawls the concept of pluralism is neutralised from the start, the aspiration of the public use of reason is to overcome pluralism as the final outcome. In fact, the deployment of rational enlightenment ethics wished for by Habermas, notwithstanding religion and metaphysics, is certainly not an environment which is favourable for pluralism and is rightly considered by Rawls as a comprehensive doctrine like the others.

---

[24]For a list of these restrictions please read my entry Ragione pubblica, in *Enciclopedia filosofica* (Milano: Bompiani, 2006), 10: 9364–9366.

[25]Even the other path taken by Rawls, i. e. the *overlapping consensus*, is clearly non dialogic.

We have seen that the public use of reason in Europe takes its starting-point from the nation states and consolidated lifestyles, themselves now deeply questioned by the legitimation which constitutional rights supply to pluralist interpretations, which, in turn, can only with difficulty be integrated into the traditional dispensations of community life. Hence, any political community should question and decide up to which point it is willing to modify its way of understanding constitutional values and their regulations. The public debate which arises should be able to transform emotional conflicts into reasonable forms. Due to the aforesaid reasons, the ethics of public debate are much more important for European societies than they are for implementing the Rawls Public Reason. And it is also more important that it be respected both in the "background culture" part of the civil society and in the "nonpublic political culture" which is part of the communication media and especially in the "broadly speaking public political culture", which is precisely that of citizens exercising their own political rights.[26]

7.5 The Ethics of Public Discourse

The ethics behind public debate do not only concern the respect for the rules of discourse, like for example those which govern the order of the debate, but firstly imply a welcoming, open attitude towards the other party and his diversity.[27] In fact, before establishing *how* to discuss, we must be aware of the common interests shared with our interlocutors as human beings and fellow citizens. In the other person we must respect the fellow citizen even when we believe his thoughts are *wrong* and the respective way of living *unsuitable*.[28]

An ethical debate starts when paying attention to the other person. It would be impossible to list what this means in detail and also probably useless since in the end respect for the other person and his opinions is not something you can teach. It is important, for instance, not to dismiss arguments (or ideas) classifying them

[26]For Rawls, instead, the proper place for Public Reason in a strict sense is represented by official, judicial and legislative courts, to the point that he claims that "in a constitutional regime with judicial review, Public Reason is the reason of the Supreme Court". J. Rawls, *Political Liberalism*, p. 231. Hence the background of the public sphere considered differs from that pertaining precisely to the European Public Reason. But it should also be considered that in Anglo-Saxon law jurisdiction is more directly tied to society. In continental Europe the public court is that of intellectuals opposed to the state reason. These are writers, philosophers, journalists or teachers and not the judicial power which is a state body. Cfr. C. Audard, "The Idea of 'Free Public Reason'", *Ratio Juris* 8, no. 1 (1995): 30–39.

[27]Here obviously as "ethics of discourse" I intend the whole attitude of the interlocutors facing one another and the virtues proper to the conversational acts. So we are not speaking of what Habermas intends by the same expression. Cfr. J. Habermas, *Moralbewusstsein und kommunikatives Handeln* (Frankfurt a.M.: Suhrkamp, 1983).

[28]Cf. J. Habermas, *Zwischen Naturalismus und Religion. Philosophische Aufsätze* (Frankfurt a.M.: Suhrkamp, 2005).

hurriedly either as obscurantist or as immoral and especially not to exclude the other from any discussion simply due to the fact that he supports a comprehensive doctrine which (either rightly or wrongly) we do not like or due to the annoyance caused by some of the ways he expresses himself or by certain distinctive clothing or behavioural signs. All of the above would add up to intolerance. It is true that the problem whether we have to tolerate intolerants might arise, however, it would still be us who stigmatise others as intolerant, whereas it would be preferable to allow everyone to speak publicly (except for those who support violence which are excluded per se) when they are willing to be subject to the analysis of reasoning.[29]

I should also add that the preconditions for respecting others depends on the attitude we have to ourselves, our opinions and our conceptions. An individual or institution can legitimately believe to know the entire truth, but when requested to enter public debate to construct the city of men, they must be ready to discuss their beliefs and also – as recently claimed by Habermas – be ready to learn from others. If the contrary were to be the case, there would be no real public debate, but only arm-wrestling, where the winner would be he who has more influence or capacity to mobilise the majority. In a deliberative democracy, the majority cannot believe it is not obligated to give good reasons for its decisions. The democratic constitutional state based on resolutions is a very form of government sensitive to the truth.

There are not only those who believe they know the truth, but also those who believe they are infallible judges of the right use of reason. And here one must remember that Public Reason is part of the realm of practical reason where the appreciation of values and the context of applying reasoning is of utmost importance. Therefore, it would be restrictive and after all dogmatic to request the use of a sole model of rationality and it is important to be open-minded to consider multiple reasons or sources of reasonableness. The only condition is that they can in some way become universal, meaning also understood and appreciated by those who do not share the same idea of the good life.

The rationalist or empiricist reductionism can be a surreptitious way to eliminate unwanted interlocutors. Even the ethical foundationalism should be left behind in public discourse which differs from philosophical debate. Practical reasoning never operates in a vacuum or in the absence of preconditions, but takes cues from pre-existing lifestyles and consolidated social practices, testing to what extent they can evolve, acknowledging needs without radically denying themselves.[30] Usually, societies develop in an uninterrupted flow and through unending internal adjustments and reinterpretations of the essential values. This obviously does not exclude the possibility that they could, for some aspects, tend to depart from their

---

[29] After all, a liberal state should also respect the rights of those who do not respect the rights and humanely treat those who have not behaved humanely. Cf. M. Ignatieff, *The Lesser Evil. Political Ethics in an Age of Terrorism* (Edinburgh: Edinburgh University Press, 2005).

[30] The mistake of modern rationalism has been to believe that our moral comprehension cannot be revisionist if it is not uncommitted. If reason were uncommitted, practical reasoning would be impossible. Cf. Ch. Taylor, "Le juste et le bien", *Revue de métaphysique et de morale* 93 (1988), p. 50.

original matrixes. The important thing is that the legitimate request for recognition is kept in mind jointly with its integration into a common vision of public life, meaning a common good.

The public debate, even when directed at recognizing practical truth as a foundation for common life, does not have a purely theoretical purpose, is not directed at establishing the superiority of a doctrine compared to another, but intends to solve the problem of communal life between people who have different beliefs, identities and convictions and, despite this, intend to live together respecting human dignity. Firstly, there are values which must be acknowledged by all, but this is not enough since the crucial question of the public debate is to re-integrate those values in history and specific contexts of a concrete political community. It is for this reason that mediation is important and it should not be confused with a negotiation of values.[31] Fundamental values are not negotiable, meaning they cannot be the subject of bargaining or compromise, but should be integrated within a life plan and this requires that there is communication between them. If just a dominant value were to exist, the public debate would become entirely unnecessary or impossible.[32] This value should prevail as the winning card. However, the essential values for safeguarding human dignity are many. The different comprehensive doctrines distinguish themselves one from the other by their varying appreciation of the importance of some values compared to others. However, if they accept being part of public debate, they implicitly accept the premise of considering the different system of values sustained by others. Participating in public debate, especially in the stage closest to the decision-making process means relinquishing irrefutable diktats. Obviously, especially in terms of a basic social culture, it is important that the different doctrines are recognized and display their true and valuable propensities. But true public debate is an arena for mutual learning and mediation.

Public debate, hence, is not directly aimed at affirming values, but at realising these in common life. Each value in itself is an irrepressible and despotic demand, but its "urbanisation" requires that it becomes a reasonable trend. Contemporary constitutionalism clearly shows the line to be followed, which, besides, amounts to the same as practical reasoning in general and as moral resolution. The values as axiological demands are constitutionalised in the form of practical principles bound to guide social action. In turn, the principles generate rules which concretise these, indicating the given action to be taken or not. Mediation is the work of principles which receive their moral capacity from the values and confer practical justification

---

[31]Cfr. M. Ivaldo, "'Valori non negoziabili' e mediazione", *Il Regno* 52, no. 4 (2007): 75–78.

[32]As for the case of Weber's ethical polytheism where the dominant values are many and incomparable.

to the rules. This is why the true subject of public debate concerns the identification and interpretation of constitutional principles.[33]

With what has been said, I would like to state that even if we follow the European way of Public Reason which does not avoid – contrary to what Rawls proposes – the question of practical truth, this does not mean that we have to look at the public debate as a philosophical one aimed at establishing the prevalence of a given conception of ethics. Fundamental meta-ethical questions as, for instance, the debate between absolutism and ethical relativism, should be left out of public debate, even though they still remain in the background of the different practical stances. Public debate is not a suitable arena for metaphysical debates and neither for confirming the validity of a meta-ethical doctrine. If it were to be used for this purpose, social-political unrest would grow without any appreciable result, even less on the theoretical level. In the public arena, the metaphysical debates tend to be reduced to an unending exchange of offences and invectives.

7.6 Civil Society and Political Society in Europe

Let's now analyse more closely a crucial matter for the present and future of the European Public Reason, meaning the public role of religion.[34] The importance of this topic is determined, amongst other things, by the fact that the European national states have historically witnessed a Christian religious presence which, through a secularisation process, has modelled the cultural foundations of society with the acceptance of non-believers as well.[35] Consequently, when the traditional settlements of public *ethos* are debated again, that will be the time to re-legitimate them, meaning to find the reasons behind their justification. So it appears that secularisation has not managed to cut all ties with its religious origins, that it does not stand on its own two legs and that – as famously noted Böckenförde – *the liberal State, secularised, lives on preconditions which it cannot by itself guarantee.* This largely explains the coexistence of two apparently contradictory social phenomena:

---

[33]"The congruential link between values-principles-rules is constitutive for the validity of law, it is an axiom which actually precedes the constitutional law laid down; it is something which comes even before the power itself to create a constitution". G. Zagrebelsky, "Diritto per: valori, princìpi o regole (a proposito della dottrina dei princìpi di Ronald Dworkin)", *Quaderni Fiorentini* 31 (2002), t. II, p. 877. Respecting this link is fundamental for public debate. It can be disregarded both when the value pretends to be asserted without being transformed into a principle and when a principle pretends to direct even when it goes against a value, or when a rule pretends to have value even when it goes against a principle. See these topics also in my article "Conflitti d'identità e conflitti di valori", *Ars interpretandi* 10 (2005): 61–96.

[34]For the topic in general cfr. my article *Il ruolo pubblico della religione nella società multicultur-ale*, in *Multiculturalismo e identità*, ed. C. Vigna and S. Zamagni, Milano, Vita e Pensiero, 2002, pp. 107–138.

[35]For the distinction between religious identity and cultural identity cf., finally, my paper "Identità culturali e religiose", *Cosmopolis* 1, no. 2 (2006): 67–74.

on the one hand, the privatisation of religion which tends to become mere devoutness without strong affiliations and, on the other, the growth of the public role of Churches and the Catholic one in particular, more tied to needs for solidarity.[36]

This explanation is more sound and solid than one uniquely preoccupied by the return of the ecclesiastic power in the political framework, a fact anyhow sometimes justified by interventions which are not always well balanced. But what especially counts is the presence of a real problem which it would be purely ideological to try and hide by resorting to the theory of conspiracy or to the philosophy of suspicion. Secularisation has proven that it is not self-sufficient and, moreover, that it has been impoverished, compared to the vitality of its religious origins.[37] This fact – well known – is one aspect of the crisis of modern times.

If, without any ideological prejudice, we come to believe that religion can last in a progressively secularising environment, then – as Habermas claims – this social phenomenon should be grasped by philosophy as a "cognitive challenge", but also more generally speaking as an existential challenge. In fact, religions enhance human aspects which escape the ethics of autonomy, such as sufferance, fragility, vulnerability, solidarity itself. The ethics of law do not exhaust that human sense and expect to communicate within the realms of interdependence and community. Hence, the direction to be followed should be the one clearly indicated recently by Habermas:

> If we intend the modernisation of public consciousness in Europe as a learning process which invests and at the same time changes religious mentalities as well as the laic ones, making the Enlightenment tradition, like the religious doctrines, reflect on their respective limits, then a different light is cast on the international clash between the large cultures and world religions.[38]

Where does the difficulty lie? Where does the main obstacle lie in this constructive dialogue? Once again, one must keep in mind the scarse propensity of European culture overall, whether it be laic or religious, to welcome pluralism and tolerate diversity. The fundamental point – in my opinion – resides in the relationship between civil and political societies. The nation state has enslaved civil society,

---

[36]Cf. S. Ferrari, "Religione, società e diritto in Europa occidentale", *Sociologia del diritto* 31, no. 2 (2004): 213–224 and bibliography hereby cited.

[37]"In a number of ways the Christian conceptions out of which modern liberalism originated remain richer and deeper than their secular offspring. For that reason - and this is my point - they continue to offer themselves as resources and clusters of clues for the modern political debate". J. Waldron, "Religious Contributions in Public Deliberation", *San Diego Law Review* 30 (1993): 846–847.

[38]J. Habermas, *Zwischen Naturalismus und Religion.*, p. 212. Amongst these limits of Enlightenment, there is the habit of criticising ideology along the lines developed by Marx, Freud and Nietzsche up to the Frankfurt School, to discredit topics ascribable to the spheres of "false consciousness" such as the religious one. In this sense, this criticism of ideology is part of the anti-pluralistic trend in Europe. Ideology, in turn, as a negative form of collective thought, should be distinguished from reasoning truly motivated by personal interest and without a universal communication capacity. To eliminate the latter from public debate, one needs to appeal to the ethics of discourse.

which is the most suitable place for the culture of diversities and has suffocated any possibility for multicultural dialogue from its inception. Consequently, the groups and individuals whose identity is not welcomed and recognised by civil society, try to request recognition from the perspective of political society and the law which taken on a symbolical function relevant and prevailing. In such a manner, however, the question does not find a solution, becomes complicated and embitters the conflict: on the one hand, in fact, since the political society is the area for determining the common good, it seems hard to reach a compromise on all matters which everyone is bound to accept as a shared value; on the other, a merely legal recognition, not supported by a corresponding attitude of the civil society, exasperates the conflicts and foreshadows the rebirth of an ethical state, even though different to the past.

Where the nation state has cut ties with pre-political restrictions, public ethics developed by political institutions, intellectuals and bureaucracy are separating from the present state in a confused manner within the core of a civil society where it has grown, even for the previously mentioned reasons, with the weight of religion and cultural identities. What is occurring in society is part of the public sphere in every way[39] and is the start of public debate, the movement of reciprocal awareness, communal life and cooperation. Without this preparation, the outcome of public debate on political society would appear as a pure and simple imposition of the "majority" within the perspective of a democracy which Dworkin has considered purely "statistical",[40] or as the result of a purely philosophical debate based on the abstract logic of constitutional rights.

Refusing to set aside the question of the truth of comprehensive doctrines does not mean – as already said – transforming the public debate into a philosophical congress, because the truth sought is "practical" and the reasons set forth are "practical", meaning understandable and acceptable by people who respect each other and intend to live together. The order of liberties comes from the heart of civil society and cannot be construed by political decisions. The European path to Public Reason cannot be taken if the reconciliation between the political and civil society is not implemented, starting from the national dimension to each single state.[41]

While for Rawlsian Public Reason – as we have seen – the problem to be solved mainly consists in setting up a political society with an eye to pluralism, European Public Reason aims towards a civil society which welcomes pluralism. But civil society is not something which can be either built or planned, being the result

---

[39]Identification between "political" and "public" is the lasting sign of a statism which dies hard Cfr. P. Hirst, *From Statism to Pluralism. Democracy, Civil Society and Global Politics* (London: UCL Press, 1997).

[40]R. Dworkin, "The Moral Reading and The Majoritarian Premise", in *Deliberative Democracy and Human Rights*, ed. H. Hongju Koh and R. C. Slye, pp. 81–115 (New Haven: Yale University Press, 1999).

[41]And here we should not generalise and learn to distinguish between the different national situations, because under this common problem lies the kaleidoscope of cultures which is the European populations.

of the spontaneous meeting (and clash) of the cultural resources of the political community. It is important to let it blossom (or wither) by itself. One can only hope that the new injection of pluralism in contemporary Europe, especially by way of a type of immigration characterised by its religious and cultural imprint, shall induce the basic culture of European societies to abandon hegemonic positions and to develop an intercultural debate respecting human rights. But signs coming from all quarters, from left and right, are not that encouraging.[42]

7.7 Public European Consciousness

So we are back to the main point which still revolves around pluralism. Shall the European public consciousness be able to keep "pluralism" in mind? To identify this consciousness with the laic processes of the Enlightenment means that this question has already been replied to in the negative. And actually – as we have suggested – managing pluralism has always been an unsolved problem for Europe. Full of strong ideological powers, Europe has lately learnt to be tolerant and not always to the fullest. European history is full of examples of persecutions for diversity, from the religious to the anti-religious. That's why the Holocaust is a key point of European history. Pluralism appeared as a scandal which had to be necessarily removed. Many methods can be used to this purpose. Stakes for the heretics, guillotine for the papists, cremators for the Jews and Siberia for dissidents are violent methods. However there are less bloody exclusion methods which are more scientific, yet nonetheless still painful. To distribute rationality patents is one of these and has a long history. We are speaking of deciding participation or not in public debate, admitting or excluding parties in it. However, if there is no respect for pluralism, the problem of Public Reason, or public use of reason in a pluralistic regime, cannot be invoked. If pluralism is eliminated, the die is cast and the matter easily solved, even though substantially betrayed.

Now, even admitting the possibility of reconciliation between civil and political society within Member States,[43] the problem of a European public consciousness still remains together with the creation of a civil society in European terms,

---

[42]Unfortunately it is significant to note that the compilation whereby Habermas describes Europe as the most laic part of the Western world does not refer in any way to the integration of immigrants which, instead, is the pride of the United States: "With the abrogation of the death penalty, liberal laws on abortion, equal legitimacy of sexual rights, equal rights to homosexual couples, absolute rejection of torture and generally speaking with the priority given to rights instead of to the collective good, for instance, national security, the European states seem to walk alone down a road which, starting from the two constitutional revolutions during the late nineteenth century, they had taken with the United States". J. Habermas, *Zwischen Naturalismus und Religion,* p. 21.

[43]For relations between one and the other please review my article "Società civile e società politica. Tra cooperazione e conflitto", in *Religione, società civile e stato: quale progetto?* ed. P. Donati and I. Colozzi, pp. 85–102 (Bologna: EDB, 2002).

something which still has a hard job in getting off the ground. This cannot start from above, meaning from a juridical-political structure like a constitution plan, which in fact encounters well-known difficulties. Integration through law (and even more through rights) cannot thrust ahead over a certain threshold beyond which the need for a public European consciousness arises.

As we have already seen, this commonality can never rise from the rubble of national cultures, but, on the contrary, should be the result of their getting together. We are not speaking of a dialogue between European citizens individually considered, but first of all between European peoples. We know that often this confrontation creates unrest. Chauvinism and the reason of state die hard. The more the European Union grows, the greater the trend to establish preferential relations between two or more states with the exclusion of others (*differentiated integration*), due to various factors like being amongst the founding countries of the Community and/or being the propellant of economic integration and of the other main concerns. We are speaking of a Europe with various degrees of velocity, of a Europe with a varying structure.[44] However, by now it is clear that a European public consciousness should stick to two conditions: respecting the variety of national cultures and not follow the forms of the national state itself. Europe cannot be a Europe for citizens if it is not first a Europe for people and it can only be a Europe for people if it is a Europe of states. In this case it wishes to follow its history which does not cancel the experience of a modern state, but remodels it.

Perhaps it is precisely in this case that we should speak of *reflective modernisation*, which recognises multiple ways of accessing the principles of modernity and in this sense represents an alternative to the *American way*.[45] However, this cultural process should come from the bottom of the civil societies of Member States, where different identities are welcomed and learn to dialogue and cooperate. This intercultural foundation is the necessary precondition to assess the different political approaches and the different legislative solutions as alternative ways to the common good, which in essence are equivalent.[46] The practical truth does not imply one sole correct answer, but a plurality of approaches and solutions whose soundness depends on multiple factors like cultural tradition, institutional history, and special social frameworks. This does not mean that any decision by political society is in itself right or acceptable. Constitutionalism invites us to consider political life as a revision process always open to corrections and reassessments. Where it implies a sacrifice of claims coming from reasonable ideas of the good

---

[44]The question has been examined by Marlene Wind, who seems to have faith in conciliating *widening* and *deepening*. Cf. M. Wind, "The European Union as a Polycentric Polity: Returning to a Neo-medieval Europe?", in *European Constitutionalism beyond the State*, ed. J. H. H. Weiler and M. Wind, pp. 103–131 (Cambridge: Cambridge University Press, 2003).

[45]Cf. U. Beck and E. Grande, *Das kosmopolitische Europa. Gesellschaft und Politik in der Zweiten Moderne* (Frankfurt a.M.: Suhrkamp, 2004).

[46]The principle of equivalence is already present in Community Law even though mainly applied to the economic and administrative field. Cf. L. Torchia, *Il governo delle differenze. Il principio di equivalenza nell'ordinamento europeo* (Bologna: Il Mulino, 2006).

life or identity requests worthy of consideration, it should be considered as an integration possibility still to be fulfilled. However, a mutual learning process between constitutional regimes is currently possible and, even though they have different institutional backgrounds, they share fundamental values and can draw benefit from the conflict of interpretations if, that is, they are not to remain prisoners of national pride and of jealous autonomy.

Chapter 8
The Social State Based on the Rule of Law in the Europe of Rights

Ernesto J. Vidal Gil

Abstract Chapter 8 shows how the idea of the social State under the rule of law has been linked to the European Union since its inception. The founding fathers (Adenaeur, de Gasperi, Monet, Schuman) set the principles of dignity, freedom, equality and solidarity as a way of freeing European citizens from fear, misery and misfortune. Welfare is, above all, a state of *justice* that defines the threshold between indecency and well-being and affirms that human development is not (only) economic growth. Jacques Maritain, whose influence on the European Union is well known, proclaimed the order that contains the three principles of the social state: the primacy of the person, the priority of *civil society* and the subsidiarity of the state. This is the great goal that the European Union must face today to fulfil the *Millenium Objectives*, reduce poverty and achieve human development, social well-being, justice and peace.

8.1 Europe, a Place of Rights: From the Common Market to the European Union

On 5 May 1949, Belgium, Denmark, France, Italy, Ireland, Luxembourg, Norway, the Netherlands, the United Kingdom and Sweden signed the Treaty of London and founded the Council of Europe, "to protect human rights, pluralist democracy and the rule of law; to promote awareness and encourage the development of Europe's cultural identity and diversity; to find common solutions to the challenges facing European society; to consolidate democratic stability in Europe by backing political, legislative and constitutional reform". The convictions of Schuman, Monnet, Adenauer and De Gasperi and their intense social commitment laid the foundations of a

E.J. Vidal Gil (✉)
Philosophy of Law, University of Valencia, Valencia, Spain
e-mail: Ernesto.Vidal@uv.es

J. Ballesteros et al. (eds.), *Globalization and Human Rights*, Ius Gentium: Comparative Perspectives on Law and Justice 13, DOI 10.1007/978-94-007-4020-4_8,

free, supportive and social Europe at the service of world peace, by founding three communities: the Coal and Steel Community (ECSC), the European Atomic Energy Community (EURATOM) and the European Economic Community (EEC).[1] For Habermas the establishment of peace and the integration of Germany into Europe, in accord with the will of the founding fathers, constituted the fundamental reason for European integration; economic union, commercial customs and excise and monetary union are means that contribute to this end, given that in the words of Lionel Jospin "Europe is much more than a market; Europe is the equivalent of a social model".[2]

The Franco-German initiative to jointly produce carbon and steel saw the birth on 9 May 1950 of a united Europe, driven by the founding fathers. The commercial and customs union was preceded by a number of profoundly significant events, such as Churchill's speech at the University of Zurich (1945) urging reconciliation between France and Germany as well as the formation of a *United States of Europe*; the actions of the *European Movement* inspired by Maritain[3] and Sturzo[4] that in the *Message to Europeans* that closed the Congress of Europe (the Hague, May 1948) demanded "a Bill of Human Rights that guarantees the freedoms of thought, of gathering, and of expression, as well as the free exercise of political opposition" and "a Tribunal of Justice capable of applying the sanctions necessary to make sure the Bill is respected"[5] and that demanded a federal, free, social and united Europe; the founding of the *Council of Europe* (1949) that from Strasburg provided the necessary impulse for the *European Meeting on the Rights of Man* (1949); the creation of the *Organization for Economic Co-operation and Development* (OECD, 1960) and the approval of the *European Social Charter* (Turin, 1961).

The *Schuman Declaration* (9 May 1950) inspired by Jean Monnet, asks not for words but "a daring and constructive act through which France acts for peace and associates with Germany". From this is born "a solidly united and strongly structured Europe"; conscious that "Europe will not be built in a single stroke nor through joint construction: it will be built through concrete actions, first creating a *de facto* solidarity" warning that "world peace cannot be safeguarded without creative efforts adequate to the dangers that threaten it. The contribution that an organized and living Europe can offer to civilization is indispensable for maintaining peaceful relations. France, which has defended a united Europe for more than 20 years, has

---

[1]Cfr. MANGAS MARTÍN, A., *Instituciones y Derecho de la Unión Europea* (Madrid: Tecnos, 2010 (8ª). DÍAZ MORENO, F., *Manual de Derecho de la Unión Europea* (Pamplona: Thomson-Aranzadi-Civitas, 2010 (5ª)). JIMENA QUESADA, L., *Dret de la Unió Europea* (València: PUV, 2006).

[2]Cfr. HABERMAS, J., "Perché l´Europa ha bisogno d´una Costituzione?", in VV.AA., *Diriti e Costituzione nell ´Unione Europea*, a cura di. G. Zagrebelsky (Roma-Bari: *Laterza*, 2003), p. 96 y ss.

[3]MARITAIN, J., *Man and the State* (Chicago University Press, 1951).

[4]STURZO, L., *Opere* en www.sturzo.it.

[5]Cfr. CARRILLO SALCEDO, J.A., *El Convenio Europeo de Derechos Humanos* (Madrid: Tecnos, 2003), p. 11 y ss.

always had serving peace as its objective. Europe was not built and there was war. Europe will not be obtained all at once nor in a joint effort: it will be had thanks to concrete actions, which in the first place create a *de facto* solidarity. The grouping of the European nations demands that the secular opposition between France and Germany be overcome, on account of which the action undertaken must first affect France and Germany".

The *Declaration* subordinated the economy to peace, "it subjected the joint French-German carbon and steel production to a common *Higher Authority,* in an organization open to the other European countries" and instituted a High Authority (Jean Monet), a Common and Parliamentary Assembly (presided over by Paul Henri Spak, De Gasperi and Schuman, successively), that "obliges the member countries and lays the first concrete foundations of a European federation indispensible for the preservation of peace" and a Common Defence inspired by Jean Monnet, through a European Community of Defence and a European army.

Nevertheless, this initiative did not prosper due to pressure from the United States, which had previously created NATO (Washington, 1949). De Gasperi proposed the creation of the European Political Community, which also was rejected in 1954 because of De Gaulle's nationalism, the split between socialists and distrust on the part of the Communists.

Sixty years have passed since 4 November 1950, when the 12 member states of the European Council signed the *Convention for the Protection of Human Rights and Fundamental Freedoms* in Rome, preceded by the signing in May 1949 of the Statue of the European Council, which in article 3 "recognises the principle of the rule of law and the principle by virtue of which any person who finds himself under its jurisdiction enjoys the human rights and the fundamental freedoms". This document, which shows the European public order of human rights,[6]

> "the governments signatory hereto, being members of the Council of Europe
>
> Considering the Universal Declaration of Human Rights proclaimed by the General Assembly of the United Nations on 10 December 1948;
>
> Considering that this Declaration aims at securing the universal and effective recognition and observance of the Rights therein declared;
>
> Considering that the aim of the Council of Europe is the achievement of greater unity between its members and that one of the methods by which that aim is to be pursued is the maintenance and further realization of human rights and fundamental freedoms;
>
> Reaffirming their profound belief in those fundamental freedoms which are the foundation of justice and peace in the world and are best maintained on the one hand by an effective political",

have agreed as follows

Article 1. **Obligation to respect human rights:**

The High Contracting Parties shall secure to everyone within their jurisdiction the rights and freedoms defined in Section I of this Convention".

[6]Ibid., p. 21.

What follows in Section I, Rights and freedoms, sets forth the catalogue of guaranteed human rights: right to life (art. 2); prohibition of torture (art. 3); prohibition on slavery and forced labour (art. 4); right to liberty and security (art. 5); right to a fair trial (art. 6); no punishment without law (art. 7); right to respect for privacy and family life; freedom of thought, conscience and religion; freedom of expression (art. 10); freedom of assembly and association (art. 11); the right to marry (art. 12); the right to an effective remedy (art. 13); prohibition of discrimination (art. 14); derogation in time of emergency (art. 15); restrictions on political activity of aliens (art. 16); prohibition of abuse of rights (art. 17); limitation on use of restrictions on rights (art. 18). Section II, European Court of Human Rights establishes the court: "to ensure the observance of the engagements undertaken by the High Contracting Parties in the Convention and the Protocols thereto, there shall be set up a European Court of Human Rights, hereinafter referred to as "the court" (art. 19). Still today, the *European Court of Human Rights* (Strasbourg) is the only court which possesses supranational jurisdiction with the power to oblige the member states to act in order to respect and protect fundamental rights. Because of this, it constitutes the highest level of protection for rights.

In the Treaty of Paris (15 February 1951), Germany, France, Italy, the Netherlands, Belgium and Luxembourg established the CECA, merged their economies and undertook the construction of an unprecedented (federal) political and social Union, whose objective was not only economic, but also political and social through the establishment of common objectives and the endowment of certain institutions with the effective abilities and powers to act in the service of a stable and lasting peace.

The Treaty of Rome, which on 25 March 1957 instituted the European Economic Community on the base of the Common Market, also founded social Europe and the Europe of rights although it lacked any explicit reference to equality (art. 7) in relation to the right of free *circulation* (art. 48) and the prohibition of all discrimination regarding employment, compensation and the other working conditions and the improvement of living conditions of workers,[7] it placed civil and political rights above social and economic ones.

The establishment of the *Common Market*, the fusion of the respective economies and the harmonizing of social policies gave rise to some unprecedented years of prosperity and material well-being, to the development, protection and safeguarding of fundamental civil, economic and cultural rights. The objectives reached give the lie to the criticism, effective though false, of it being the Europe of mercenaries and show that the intention of its creators far exceeded the mere establishment of a commercial union.

In the Preamble, the signatories "resolved to ensure the economic and social progress of their countries by common action to eliminate the barriers which

---

[7]Cfr. JIMENA QUESADA, L., "La Carta Social Europea y la Unión Europea," *Revista Europea de Derechos Fundamentales*, 13 (2009): 394.

divide Europe, affirming as the essential objective of their efforts the constant improvements of the living and working conditions of their peoples, resolved by thus pooling their resources to preserve and strengthen peace and liberty, and calling upon the other peoples of Europe who share their ideal to join in their efforts".

Lastly, the Preamble of the *European Social Charter* (ESC, Turin 1961, revised in 1996), "considering that the aim of the Council of Europe is the achievement of greater unity between its members for the purpose of safeguarding and realizing the ideals and principles which are their common heritage and of facilitating their economic and social progress, in particular by the maintenance and further realization of human rights and fundamental freedoms", incorporates social rights into the EU. Before the emphasis on free circulation (of merchandise, not of persons) and of safety (Schengen), the European Social Charter (ESC), an emblematic manifestation of the social state of law, shows that social policy is no longer the "poor relation" of the EU.[8] Rightly did the President of the European Committee of social rights affirm that ESC is constructed within the European social democracy pact *par excellence,* which, by the imperative of the principle of the indivisibility of civil and political rights and social and economic rights, must cease to be the "poor relation" of the European Human Rights Convention.[9]

The achievement of peace and the developments that in time will be called "human", of economic well-being, of social justice, of freedom, of equality and of solidarity, respect for rights and the enforceability of *positive* rights, were 60 years ago and remain in 2010, the goals that propel and guide the European Union. For this reason, when faced with the commercial and economic criticisms of those who only see the *Europe of commerce*, motivated exclusively by the attainment of benefit no matter by what means nor for what ends, by efficiency at the expense of justice and by the suppression of every kind of obstacle that impede the free functioning of the market (deregulation, minimal state), it is fitting to recall the aims and objectives that the signatories declare in the Preamble of the Treaty: "to ensure the economic and social progress of their countries by common action to eliminate the barriers which divide Europe, affirming as the essential objective of their efforts the constant improvements of the living and working conditions of their peoples, resolved by thus pooling their resources to preserve and strengthen peace and liberty, and calling upon the other peoples of Europe who share their ideal to join in their efforts".

---

[8]Cfr. JIMENA QUESADA, L., "Retos pendientes del Estado social español: en especial, la ratificación de la Carta Social Europea revisada de 1996", *Nuevas políticaspúblicas* núm. 2, (2006): 41–55.

[9]Cfr. JIMENA QUESADA, L., "La Carta Social Europea y la Unión Europea" *art. cit.* p. 391.

8.2 The Social State of Law: Rights and Positive Duties

The *Reichtstadt*, is also a European invention,[10] which arose in Germany under the impulse of Kant, Von Mohl and Von Humboldt and within the *iuspublicista* dogmatic (Welckel, 1813)[11] to limit the arbitrariness of the absolute state.

The rule of law demands the rule of *leges, per leges et sub leges* before the rule of *reges*. Aristotle (Politics 1286, a) postulates the *isegory* and *isonomy,* and defines a state of reason and power submitted to the law that excludes arbitrariness. For Max Weber, legality, certainty and security[12] constitute the rational legal state and produce the rational legitimization of law that is imposed coactively and demands obedience (*ratione imperio*) because it contains the *imperium rationis.* The *Reichtstadt* limits its own powers and guarantees (Jellinek) the (subjective) rights of the citizen.

The *laissez faire, laissez passer* liberal-constitutional state ascribes to the law the tutelage and repression of order (*to watch over and punish, night-watchman*); the minimal state is incompatible with the social state that before neutrality proposes impartiality, promotes positive discrimination (actions of inverse discrimination and/or positive affirmation, the use and also the alternative abuse of the law), assumes *premial* law and positive sanctions (Bobbio) and confers upon law the active and positive function of promoting real and effective equality. Before the *natural lottery* and the *invisible hand* the social state guarantees the *minimums of justice*: entrance, feeding, health, housing, education to every citizen as a political right and not as beneficence.[13]

The social conquests of the end of the nineteenth century, driven by the working class and by the incipient social doctrine of the Church configured the new *social* law that limited the abuses of the *individual* law. At the height of the industrial revolution, the laws of Bismarck created a precarious model of social paternalistic and "benefactor-style" protection to silence the claims of the unions that in the Weimar Republic consolidated the *social state of law* (Heller).

---

[10]Cfr. MATTEUCCI, N., "Stato", in (ID), *Lo Stato Moderno. Lessico e percorsi*, pp. 15–79 (Bologna: Il Mulino, 1993).

[11]Cfr. BALLESTEROS, J., "Los principios básicos de la Constitución (I): el Estado social y democrático de Derecho", en VV. AA., *Estudios sobre la Constitución Española de 1978* (Valencia: Universidad de Valencia, 1980), p. 37 and ff.

[12]Cfr. LÓPEZ DE OÑATE, F., La *certezza del Diritto*. Nuova edizione riveduta. In aggiunta, saggi di G. Capograssi. P. Calamandrei, F. Carnelutti e P. Fedele (Giuffrè: Milano, 1968). PÉREZ LUÑO, A. E., *La seguridad jurídica* (Barcelona: Ariel, 1991). MARTÍN MORÓN, Mª T., *El deber general de conocimiento de la norma y su proyección enel ámbito contracutal* (Pamplona: Thomson REuters Aranzadi, 2010).

[13]Cfr. REGONINI, G., "Estado de Bienestar", en VV. AA., Bobbio, N., MattEucci, N., Pasquino, G., *Diccionario de Política.* Traducción de R. Crisafio (México: Siglo Veintiuno de España, 1982), pp. 551–556.

Its precedents are found in Vives, who in *De subventione pauperum*,[14] denounced social inequalities and poverty. The *Cameralistica* guaranteed *pax et tranquilitas*; Von Mohl joined liberalism with the social state and incorporated economic, social and cultural development. Von Justi linked the *Sicherpolizei* or security police with the *Wholfahrtspolizei* or (social) well-being policy, since there is not security in the state if poverty persists. In the middle of the nineteenth century, Von Stein noted the duty of the administration to attend to social needs.[15]

The imposition of limits on the exercise of rights such as the abuse of a right, good faith and fraud; the objective responsibility and the limitations to the right to property, previously absolute and now subordinate to their *social function*, (art. 135 Weimar Constitution, 14 Grundgesetz, 42 Const. It, 33C. Esp), prove that the law is not reduced to *keeping watch and punishing*, that the (liberal) state cannot remain indifferent in front of injustice and inequality and that it must fulfil some new *positive* tasks of promotion and redistribution from where rise those rights whose foundation is solidarity (the *good Samaritan*) and the *positive rights* considered in art. 21 of the unsigned Jacobin *Constitution* of 1793 as a *sacred debt to society*.

The Social Service State or Welfare State, preceded by the New Deal (Roosevelt, 1933) and the Social Security Act (1935) and by the social laws of New Zealand (1938), were consolidated in the *Beveridge Report, Social Insurance and Allied Services* (1942) that, faced with the disasters of war and totalitarianisms, and the authoritarian, patriarchal, assistential and paternalistic systems (Bismarck) and those systems of philanthropy, benevolence and private assistance and volunteers from liberal society, guaranteed the right of all citizens to health (*National Health Service Act*, 1946), to education and to the satisfaction of basic needs (*Family Allowances Act*, 1945, *National Assistance* 1948) from a universalist and supportive basis that included *ex ante* provisions (*National Insurance Act*, 1946) and conferred a public and social character on the *ex post provisions*, until then subject to the regime of private law (civil and mercantile) insurance.

The *European Recovery Program* (Marshall Plan), the CECA Treaty (Paris 1951), preceded by the Schuman Declaration, (9 May 1950) and by the CECA and EURATOM Treaties (25 March 1957) inspired by the founding fathers of the European Union (Adenaeur, De Gasperi, Monnet, Schuman) who established the CEE and are the seed of the European Union, confirm the irreversible nature of the social state of law that has strengthened democracy and has promoted economic growth and the redistribution of wealth; social well-being (health, education, pensions) and human development (quality of life, capabilities).

The social state generates security, transmits confidence, provides freedom from fear (four freedoms: freedom of expression, freedom of religion, freedom from want, freedom from fear, Roosevelt, Annual Message to Congress, 6 January 1941), and creates the conditions for the individual to develop his autonomy. It is a state

[14]MONZÓN, A., *El Derecho en J. L. Vives* (València: Universitat de València, 1987).

[15]Cfr. SOTELO, I., *El Estado social, antecedentes, origen desarrollo y declive* (Madrid: Trotta, 2010), p. 145 y ss.

committed to liberty and equality and also to guaranteeing dignity for all, and whose (*positive*) purpose is to remove inequality, redistribute wealth and promote dignity for all. As opposed to the *Libertarian model* which only admits the rights of the first generation, we have the *Utilitarian*, which knows what the markets can and cannot do, respects the rights of first and second generation, regulates the market and redistributes the benefits.[16] Today the Nordic countries, the European Union, Australia and New Zealand show that it is possible to join freedom, equality and solidarity. The social state must guarantee a *decent minimum*, which is a duty of justice and overcome *economic* reductionism so as not to confuse, the protection of basic rights with the satisfaction of infinite desires.[17]

The crash of '29 and the crisis of the 1930s showed with Keynes that the market is incapable of establishing growth and reaching full employment. On the contrary, the crisis generates incertitude, a lack of confidence and fear. Because of this, the social state of law substitutes the *incertitude of divine providence with the certainty of state action* (Rosanvallon) and represents the triumph of equality over selfishness; before the market economy the social state fights against poverty (the role of Robin Hood) and realizes an efficient redistribution (the role of the social provider). For Forsthoff, the *Daseinnvorsorge*, or existential endeavour, shows that the *social* state is incompatible with the *liberal* state because the social state denounces the fictitious liberal neutrality and is indifferent to the necessities and rights of the weakest and most vulnerable and those that it excludes and marginalizes[18]; because of this it intervenes *a priori*, even at the cost of rights and guarantees (*the law of the weakest*, Ferrajoli), it takes on the *positive* duties, proclaims social rights and solidarity rights, carries out the task of promotion and promotes *premial* law, *affirmative action* and policies of *reverse discrimination.*

The grounds of law and of rights is no longer as in Malthus' *Parable of the Banquet*, the sacralisation of the market, exclusion and marginalization, nor structural, institutional, economic, social and political violence, but rather their elimination. The *social state* reacts against the presumably natural and inexorable laws that guided by the *invisible hand* rule the market, to guarantee, against the unjustifiable privileges of the rich and powerful, that rights be respected, the fulfilment of *positive* duties, the realization of *capabilities* and the satisfaction of basic needs.

In comparison with the populist, electoralist, and "clientelist" connotations of the *well-being* state and the ethical justification for the social state, the *Welfare* state *does, if* rights are *taken seriously*, consist in fulfilling the ethical demands of dignity,

---

[16]Cfr. LUKES, S., "Five Fables About Human Rights", in *On Human Rights: The Oxford Amnesty Lectures 1993*, ed. Shute, S. and Hurley, S, pp. 19–40 (New York: Basic Books, 1993).

[17]Cfr. CORTINA, A., *Ciudadanos del mundo. Hacia una teoría de la ciudadanía* (Madrid: Alianza, 1999), p. 86 and ff.

[18]Cfr. FORSTHOFF, E., "Concepto y esencia del Estado Social de Derecho", in VV.AA., ABENDROTH, W., FORSTHOFF, E., DOEHRIG, K., *El Estado social*. Trad. José Puente Egido (Madrid: CEC), pp. 69–106. The note, on p. 86.

liberty and equality and solidarity, guaranteeing rights and demanding (positive) duties, respecting the individual, empowering his capabilities and increasing the autonomy, freedom and independence of its citizens. For this reason it is a state of *justice* and not of *welfare* or *well-being* (Cortina). Rightly does Sen write that *Welfare is one of the great achievements of European civilisation, one of the great contributions of Europe to the world.*[19]

8.3 Civil Society and State

We have seen that Europe is built upon the roots of Christian humanism, of the Enlightenment, of democratic socialism, of liberalism and personalism, that from the base of tolerance and pluralism, proclaim dignity and freedom, equality, and solidarity. The *European Common Constitutional Law* (Häberle) orders and limits property and the market and state, subordinated to social interest and the Common Good; it permits (fair) trade and free trade, the *ius migrationis* and *ius comunicationis*, which other illustrious Europeans anticipated, the fathers of international law and law of the people (Grotius, Francisco de Vitoria and Kant), to contribute to peace, strengthen the relations of peaceful cooperation between peoples, promote *hospitable and cosmopolitan* law and prepare the *Weltrepublik*. For this reason, the European legislation committed to peace, with respect for (human and fundamental) rights and to the social state, rejects the current deregulation and abuses, and accordingly, regulates and keeps watch over the markets so that they may fulfil their social function; it demands the fulfilment of *positive* rights and duties, strengthens civil society and reinforces the principle of subsidiarity before the strong states submitted to the rule of the constitution and the law.

Civil society, (vs. Uncivil society), *civil economy* (vs. *Political* economy and *homo oeconomicus*) and *civil humanism*, which are some of the presuppositions that inspired European law, arise in the cities of Tuscany and Umbria of the fifteenth century and maintain *civil, horizontal and communitarian* relations based on reciprocity, gratuitousness and solidarity that are opposed to "economicism", (to think is to weigh, to measure, to calculate; time is gold); to the modern individualism (*homo fäber*) and social Darwinism[20] and constitute a frame of reference to understand the construction of the European social state. Committed to the construction of the secular civil order, the first treatises of economy of monastic inspiration translate the *rule* into civil order, they order time and work (*ora et labora*) in the city, submit economic and commercial relations to the *common good* and the *public happiness* of the citizens and create *civility*. These criteria are opposed to proponents of the Anglo-Saxon trade treatises in which, during the 6 days of the week the Bible of traders was the Great Book and only on Sunday was the Great

[19]Cfr. CONILL, J., *Horizontes de economía ética* (Madrid: Tecnos, 2004), p. 167.

[20]Cfr. BALLESTEROS, J., *Sobre el sentido del Derecho*, 1ª ed. (Madrid: Tecnos, 1984), pp. 19–43.

Book the Bible. Not by chance do the *Montes de Piedad* constitute a*n* alternative bank to the market that, anticipating the current micro credits, share out, with equality, solidarity, reciprocity and credit for the poorest and most vulnerable of the city, to re-establish the city damaged by their exclusion.[21]

The *civil society* is opposed to the (uncivil) state. The *night of civil society* (Zamangi) shows that the decline of civility is correlative to the rupture between ethics and politics, economy and law (Machiaveli, Hobbes), which distinguishes between *private vices* and *public virtues* (Mandeville) and separates *private* ethics from *public* ethics (Mill), the ethics of *conviction* and that of *responsibility* (Weber). *Civil humanism* sustained in reciprocity and in gratuitousness is opposed to the perpetual state of violence and war (*bellum omnium contra omnia*) where men live *more ferarum*, whose only exit is the pact that constitutes the political society, demands absolute obedience, proclaim civil religion (Rousseau), makes of the citizen a believer (Hegel) and submits all to the state (Hobbes).

In contrast, the *civil society* reappears with the triumph of the liberal revolution (Locke) which recognizes rights, develops the market, diffuses civility, separates the Church and state and allows for the right to resist (Locke, the Spanish school of the law of the people, but not Kant); with the division of powers that limit the power of the state (Montiesqieu) and with the *Republican Constitution* that guarantees *freedom inasmuch as one is man, equality inasmuch as one is a subject, and independence inasmuch as one is a citizen* and makes *unsociable sociability* possible (Kant).[22]

In comparison with the crude *oriental* societies subject to powerful and arbitrary tyrants highlighted by Ferguson in his *Essay on the History of Civil Society* (1767) and Hegel's *despot*, the *civil society* that is the society not directly submitted to state action, shows a tendency to become associated with and to surpass the individualism that Tocquevill praised in his *Democracy in America.* But it is no longer that *bürgerliche gesellschaft*, individualist, competitive and selfish in which *each one is an end in himself and the others are nothing for him* (Hegel), but the *zivilgesellschaft* or *civic society*, that configures a realm of recognition and cooperation where persons and not selfish individuals share the projects for a *happy and good life*, promote civic commitment, establish support networks and elevate the moral temple of society.[23] It is a *civil and civilised society* composed of free and autonomous citizens who demand a limited state that respects their autonomy and independence and favours peaceful competition, general prosperity and amiable

[21]Cfr. BRUNI, L. y ZAMAGNI, S., *Economía civile. Efficienza, equità, felicità publica* (Bologna: Il Mulino, 2004).

[22]Cfr. PÉREZ LUÑO, A. E., "El papel de Kant en la formación histórica de los derechos humanos" en VV.AA. *Historia de los derechos fundamentales*. Tomo II, vol. II La *filosofía de los derechos humanos*, pp. 451–483 (Madrid: Dykinson, 2001). CONTRERAS, J., "La libertad en el pensamiento de Kant", op.cit., pp. 483–577.

[23]CORTINA, A., "Sociedad Civil", en VV. AA., *10 palabras clave en filosofía política* (Adela Cortina dra) (Estella: Verbo Divino, 1998), pp. 352–387.

social relations, being governed by a minimal civil – and not military – power.[24] Currently, the *civil society* includes a framework of socio-political institutions that include a limited government (or state), under the rule of law; a union of social institutions such as the markets and associations based on voluntary agreements between autonomous agents and a public sphere where these agents debate among themselves and with the state about matters of public interest and are committed in public activities".[25]

8.4 Political Economy, Social Market Economy, Civil Economy

The supremacy of the state and of the *political* over *civil society*, determined the rule of the political economy built upon selfish, and presumably rational interest, corrected by the *invisible hand* and sustained in the fictitious neutrality and by the real abstention of the state.

With the exception of Stuart Mill, who proposed an egalitarian, progressive and social liberalism, Liberalism gave priority to efficiency, utility, and (presumably) rational selfishness and individual benefit, and drew up the *great division* between the private sphere where the market, benefit and production moderated by compassion, philanthropy, charity and the public, were reduced to the *minimal state.*

The *social market economy (Soziale Marktwirtschaft)* explicitly present in art. I-3 of the TEU, is another European invention that, together with the *ordoliberalismus* was developed in the 1930s as a reaction to totalitarianism and war, and aroused political and social consensus, to the point of being taken on by the Christian Democrats, Social Democrats (Bad Godesbers, 1959) and by the Liberals, for its capacity and to put brakes and limits on the markets and the state and at the same time, to recognize individual freedom and private initiative and to sustain a limited but strong and demanding state. The *Sme* confers on the state the function of guaranteeing economic order and ensuring prosperity for all and of joining free competition with solidarity and subsidiarity.

Since its implementation (Erhardt, 1945), the *Sme* ordered and limited the market, sustained German growth, mediated between Capitalist Liberalism and state interventionism, diffused social congestion (*Miltbestimmung*), joined individualism, solidarity and subsidiarity and granted the state the function of putting order in the economy, guaranteeing rights (individual, economic, social and cultural) and redistributing riches through taxes. The *Sme* that has consolidated the *Welfare* state and has married economic growth, social well-being and the guarantee of rights has

---

[24]Cfr. GINER, S., "Sociedad civil", in VV. AA., *Enciclopedia Iberoamericana de Filosofía.* ed. E. Díaz and A. Ruiz Miguel. *Filosofía Política. II. Teoría del Estado*, pp. 117–145 (Madrid: Trotta, 1996).

[25]Cfr. PÉREZ DÍAZ, V., *La primacía de la sociedad civil* (Madrid: Alianza, 1993), p. 77 y ss.

ensured the continuity of social policies, and has maintained its achievements until the present day.

In the face of current deregulation and unbridled capitalism, the *Sme* reaffirms the necessity of joining the economy and ethics, the individual and the social; of recovering civic values and of building Europe, in accordance with the *Social European Charter* and the *European Charter of Fundamental Rights,* starting from freedom, equality, justice and solidarity. The rupture between economic growth and development, the *Easterlin paradox* that demonstrates that economic growth creates instability, does not increase well-being nor happiness and damages the environment[26] and the eruption of human development (Sen), confirm that the leaders in the *Human Development Index* have applied the *Sme.*

The *civil economy* re-enacts this bond between ethics and economy and configures itself as an alternative to overcome the present crisis. Its objective is to join three principles in the social system: that of *efficiency which* rules the exchange of goods; of *equity* that orients the distribution of wealth and of *reciprocity which* presides over the relations of gratuity and fraternity. While it pursues the total good, the civil economy seeks the *common good*; while production maximizes benefit, efficiency and the utility of the *homo oeconomicus,* distribution brings about reciprocity, compassion and gratuitousness, from whence arise the *positive duties* that require the intervention of the state and of the *civil society* bypassing the former state-given, paternal and total *welfare state*.[27] In contrast, the *new welfare* is subsidiary, respects individual capabilities, is integrated into the networks of both the *civil* and *private societies*, and promotes the *Third Sector* where *non-profit organizations*, foundations, and the volunteer core *care for* basic necessities, *dependence* and the social services that were previously handled by the family and/or the (social) state and are extended as far as international cooperation, human development, peace and global justice.

8.5 The New Citizenship: The Third Sector and the New Welfare

Citizenship, polités, civis, citoyen, indicate *belonging* to the polis.[28] Nevertheless, even today, that democracy that is *isonomy* and *isegory*, sustains, against the universality of rights, the unjustifiable *privilege* of those citizens who appropriate for themselves the rights and *excludes* those who do not possess this *status*. While

---

[26]BRUNI, L., *La ferita dell´altro* (Trento: Il Margine, 2007).

[27]Cfr. BRUNI, L. Y ZAMAGNI, S., edd. *Dizionario de economía civile* (Roma: Cittá Nuova, 2009).

[28]Cfr. CORTINA, A., *Ciudadanos del mundo. Hacia una teoría de la ciudadanía*, cit, caps. II y III. (ID) *Alianza y Contrato. Política, Ëtica y Religión* (Trotta: Madrid, 2001), p. 55 y ss. PÉREZ LUÑO, A. E., *Ciberciudadani@ o ciudadaní@. com* (Barcelona: Gedisa, 2004).

for political philosophy, citizenship defines the *status* of the citizens and considers exclusion as *natural*, even today not every human being is a person nor possesses the basic rights, moral and juridical philosophy dependent on the recognition of the other and on responsibility, accenting the rights and *positive duties* founded on solidarity.

Liberal citizenship that limits the subject of rights to *males, property owners, and the Bourgeoisie*, and *social citizenship* (Marshall) that only *protects* the citizens, exclude those who need the most protection and care. A *just citizenship* is impossible and a *cosmopolitan* citizenship, impossible. Rightly did Ferrajoli recall that the condition of citizenship in our wealthy countries represents the ultimate class privilege; the ultimate factor of exclusion and the ultimate contradiction do not agree with the universality of the fundamental rights.

If in antiquity the *natural end* of man was to participate in social and political life; if to abandon the state of nature, freeing oneself from the *sordid and cruel* existence and building the state (*absolute*, – Hobbes –, *liberal* -Locke, Montesquieu, Kant, Smith and Tocqueville – and *total* – Rousseau, Hegel-) was a moral duty, at present, when faced with the excess of citizenship (liberal and/or republican) and the lack of civility (social), we must respect the basic rights of the person, free *civil society* from its subjection to the state and rebuild a social state, submitted to the rule of law, strong, just, and subsidiary. While the *liberals* defend freedom *as non-interference* but are incapable of protecting those who most need it, the *republicans* promote *public virtues* and hide private ones (Walzer), they remind us of the ties of patriotism and *love for the fatherland*, they proclaim *free living* and exalt *non-domination* (Pettit, Virolli). But they cannot elude its perverse effects such as authoritarianism, Jacobinism, paternalism, the dissolution of the private, the subordination of rights to common utility and to the *tyranny of the majority*, which produce an oxymoron: the *servant citizen.*

The *new citizenship*, inherent in the social state, recognizes the person as a subject of rights and promotes his civic participation. It is neither *homogenous* nor *simple* but *complex* (Cortina), values the differences in *capabilities*, religious beliefs and political sensibilities, it resides in institutions and expresses the recognition of the other with equality, reciprocity and solidarity. It creates networks of support and cooperation that go beyond utility and sympathy and configures new relationships such as generosity, gratitude, respect, consolation, and care for the destitute. Its reference is no longer economic or political (*lib/lab*), but *relational, civic and social.*[29] As opposed to impersonal, administrative, *societal*, corporate and vertical relations, the interpersonal, horizontal, *communitarian* and open relations arise from the recognition of the other, they promote *communities of care* that give meaning to life, respect the individuality of the subject and promote his *capabilities* and

[29]Cfr. DONATI, P., "Ciudadanía lib/lab (≪Tercera Víax≫) *versus* ciudadanía societaria (Civilización)", en VV. AA. *Las Terceras Vías*, ed. J. Pérez Adán, pp. 49–82 (Madrid: Ediciones Internacionales Universitarias, 2001).

replace the incapacity of governmental agencies to care for and love, where no one is excluded.[30]

In comparison with *political* relations (power, *domination), societal* relations (market), and *administrative* relations (bureaucracy), *communitary* relations proclaim the *compassionate reason* from whence solidarity flows. As opposed to the closed, selfish and mercantile solidarity that excludes those who *are not* like us, a *fraternal and open* solidarity recognizes individuals, respects their differences and redistributes justly and efficiently the scarce means to those who have none. In comparison with the mercantile exchange of *societal* relations (*do tu des*) that treat *brothers as strangers* (Weber), *communitary* relations recall the principle of state subsidiarity, uncover the fictitious liberal neutrality and oblige it to act as an impartial agent that must guarantee the *minimums of justice* and promote citizens that are not *servants* but autonomous and interdependent beings. The reciprocal recognition of you as other, equal and different and the imperative that commands that one treat the other always as an end and never as a means (the new *golden rule*) may lay the foundations for interpersonal, civic and social relations ruled by the principles of dignity, liberty, equality, reciprocity and solidarity that does not distinguish[31] between *Jew or Greek, Roman or Gentile.*

The *Third Sector* configures a complex of institutions situated between the state and the market. Faced with the *market or state* dichotomy, the Third Sector, the volunteer sector and the NPO (*Non-Profit Organizaitoins*) is positioned in the United States against the (*non-profit*) market or in Europe, against the state (*NGOs*).[32] The *social* is the nucleus of the *Third Sector, spontaneous, voluntary, and free*, generating social values and virtues (equality, justice, reciprocity, solidarity, participation, commitment, loyalty, pluralism, tolerance), that enrich the moral quality of the subject and social decency, and constitute the *private social sector.* The *NPOs* share the *ethos* of the civil economy, value *relational goods* and confer primacy on the *civil society as opposed to* the state, whose ineptitude to redistribute fairly the scarce goods, to value the capabilities and to generate reciprocal relations demand subsidiarity; because of this, they do not compete with the state but, in a subsidiary manner, make up for its limitations.

While the *Welfare state* only attends to its citizens, the *Welfare society* occupies itself with those excluded (from society, from education and from work). As opposed to "clientelism", paternalism and the clientary domination of the *Welfare state,* the *Welfare society* reminds us that freedom is *not domination* (Pettit), if it *takes rights seriously* (Rawls, Dworkin) and accentuates the *capabilities* (Seen) that people have in conformity with their dignity: economic opportunities, the capacity for social relations, for access to education, for understanding politics, for accomplishing valuable projects and of manifesting sentiments that will increase

---

[30]VV. AA*, Glosario para una sociedad intercultural*, coord. J. Conill (Valencia: Bancaja, 2003).

[31]Cfr. COTTA, S., *Il Diritto nella essistenza. Linee di ontofonomenologia giuridica.* (Milano, Giuffrè, 1985), pp. 67–89.

[32]Cfr. BASSI, V., "Non profit", *Dizionario di economia civile*, cit. pp. 627–634.

social integration and respect for those who are different and will promote well-being, will promote justice and will seem like human development. In other words income, literacy, access to education and health care and to an adequate lifestyle and life expectancy.[33] It is not an "economist" *Welfare*, utilitarian, efficient, paternalistic, clientary and authoritarian that humiliates its subjects and generates dependency, but a *habilitating Welfare* of the freedoms and capabilities that unite solidarity, fraternity, and reciprocity with subsidiarity.

The *Welfare network* that unites public, private, social and market sectors, recognizes institutionally the action of primary groups like the family, excluded by the state's influence and is widened to non-subjects who are traditionally excluded, such as immigrants, the *inefficient* and those who *are not WASP (white, Anglo-Saxon and Protestant). From* Titmuss's classic typology that distinguishes between a model that is *residual or liberal, institutional or social-democratic* and *radical or socialistic*, Esping-Andersen sustains that a policy of well-being is a combination of a mixed production of well-being between the state, family, market and *civil society* that varies depending on of the level of demercantilisation and of social stratification and gives rise to three models of well-being: the *social-democratic or Scandinavian* (high rate of demercantilisation and universality); the *corporate-conservative* (with a moderate rate of demercantilisation and familiar and corporate solidarity) and the *liberal* (with low demercantilisation and a low level of universalism). Leibfried presents four models: the *Anglo-Saxon or residual,* the *Bismarckian or institutional,* the *Scandinavian-modern* and the *Latin Mediterranean* and Ferrera distinguishes between the *Anglo-Saxon,* the *Bismarkian*, the *Scandinavian* and the *South of Europe.*

8.6 The Subsidiarity of the State

The principle of the subsidiarity of the state reappears today with unusual strength. In brief, it says that the state must abstain from carrying out what individuals and smaller communities can and should do.[34] The principle, formulated by Leo XIII and taken up by the social doctrine of the Church is opposed to liberalism and totalitarianism and mediates between selfish and "economicist" individualism of *homo economicus* the collectivist and holistic "statalism"; because of its opposition to individualism and economicism it rejects the *minimal state*; because of its defence of the individual and of the *open society* (Popper) it rejects the *total state.*

The subsidiarity that is upheld in the individual and sustains the primacy of the person and of the *civil society* before the state, demands a state submitted to the rule

[33]Cfr. CONILL, J., "Capacidades humanas" en VV.AA., *Glosario para una sociedad intercultural*, coord. J. Conill (Valencia: Bancaja, 2003), pp. 30–34.

[34]Cfr. PIZZOLATO, F., Sussidiarietà, en VV. AA. *Dizionario di economía civile*, cit., pp. 745–752. VIOLA, F., "Luci e ombre del principio di sussidiarietà", *Ragion Prattica* 32 (2009): 107–130.

of the law, *social, strong and limited*; a social state that oversees the markets and redistributes with justice and equity; a *strong* state that shares responsibilities with the *civil society*, that overcomes the "statalism", economicism and paternalism of traditional *Welfare*, that recognizes freedom, fights for equality, promotes solidarity and undertakes new forms of cooperation with individuals and with the *civil society*; that evolves the *Welfare state* into the *Welfare society*, and, lastly, a *limited* state whose power is submitted to the law.

Subsidiarity adopts a personalist anthroplogy that, without detracting from the individuality of the subject (*individuae substantia*) reaffirms the person as a relational being, finite and limited, cooperative, not competitive, who constructs his identity with the other and fulfils himself fully in reciprocal recognition and in the encounter with the other; who seeks the *common good*, recognizes the communitary bond and promotes personal and social commitment. The subsidiarity of the state recognizes capabilities and respects the ability of subjects to design their lives to yield the greatest from their capacities; it respects the *open society*, rejects organicism and collectivism, defends the liberty of the individual and when faced with communitarian excesses, points out the risk of the greedy institutions (Kymlica, Walzer).

Civility, reciprocity, solidarity, gratuity and fraternity are the values that define its development. From here arises the paradox, for it calls for a strong state and at the same time a limited one that will recognize the person and his inalienable rights and will fight for liberty, equality and solidarity, that commits itself to the *common good* and is able to fairly redistribute scarce goods; that will establish the minimums of justice and, – in this we see the subsidiarity – may allow each member to seek their own greatest happiness. The principle of subsidiarity orders and limits the market; it opposes deregulation and promotes the *civil economy* that orders and guides the market and is sustained by civility, confidence, reciprocity and gratuity and by relations defined by the search for the *common good.*

8.7 From Maastricht to Lisbon: The Social State in the European Union

A *communis opinio* points out that the process of the EU passes through three phases: free exchange, union customs and common market. The Single European Act (1986) completed the economic union with economic and social cohesion (establishing the bases for cooperation – FEDER- and simultaneously initiating the path towards a single currency) and with political cooperation, establishing for the first time a communitarian foreign policy (art. 30). Events in the East and the fall of the Berlin Wall (1989) hastened the political agenda which saw the Council of Dublin (1990) urge drawing up of a Treaty of the Union.

The Treaty of European Union (TEU, Maastricht, 7 February 1992), constitutes the most important reform of the EEC because it *created* the *European Union*, changed the name of the European Economic Community to the *European Union*, introduced new forms of cooperation between the governments of the member states (defence, justice, interior) and created a new structure with "three pillars": European Communities, Common Foreign and Security Policy, (CFSP), and Justice and Home Affairs (JHA).

Art. 2 enumerates the purposes of the European Community: sustainable and non inflationary growth, convergence of economic results, a high level of employment and social protection; an elevation of the level and quality of life; economic and social cohesion and solidarity between the member states that will be achieved through the establishment of a Common Market and of an Economic and Monetary Union.

Finally, as A. Mangas points out, through the reform instituted by the Treatise of Amsterdam, the Union commits itself to the democratic principle and the principle of respect for human rights. Art. 6 says, "the Union is based on the principles of liberty and democracy, respect for human rights and for fundamental freedoms and on the rule of law, principles that are common to the member states". From here arises the EU's commitment to respecting the fundamental rights that incorporate the jurisprudence of the TEDH (Nold case), consecrates the normative patterns that every communitarian act must respect under pain of nullity, guarantees the European Convention for the protection of the rights of man and incorporates the general principles of communitarian law that substitute some of those laid out in this regard in Maastricht. The reference to values works in two directions: external, since according to art. 49 any European state that respects the values set forth in section 1 of art. 6 can request entrance as a member of the Union, and internal, since respect for the principles of freedom and democracy is a requirement for remaining in the Union. Art. 7 establishes the sanctions, which do not include expulsion but rather the suspension of certain rights of the member state.[35]

The Treaty of Nice (2001) was incapable of conferring normative value to the Charter of the fundamental rights of the Union, which nevertheless proclaimed a political agreement. The Council of Laeken approved the *Convention* that drew up the *Charter* that was presented in the Council of Thessaloniki and signed in Rome on 29 October 2006.

On 1 December 2009, after overcoming the last obstacle set by some member states for its ratification (France, Ireland, Denmark, Germany, the Czech Republic and Poland), the *Treaty on European Union* (TEU) entered into force, as well as the *Charter of Fundamental Rights of the European Union which* consolidated European common constitutional law (Häberle) regarding the development of some common principles. The Preamble of the TEU (consolidated version), drawn up as part of the Treaty of Lisbon (2007), states that the Heads of State and Plenipotentiaries:

> "RESOLVED to mark a new stage in the process of European integration undertaken with the establishment of the European Communities,

[35]MANGAS MARTÍN, A., *Instituciones y Derecho de la Unión Europea.* cit, p. 67 and ff.

DRAWING INSPIRATION from the cultural, religious and humanist inheritance of Europe, from which have developed the universal values of the inviolable and inalienable rights of the human person, freedom, democracy, equality and the rule of law,
CONFIRMING their attachment to the principles of liberty, democracy and respect for human rights and fundamental freedoms and of the rule of law,
CONFIRMING their attachment to fundamental social rights as defined in the European Social Charter signed at Turin on 18 October 1961 and in the 1989 Community Charter of the Fundamental Social Rights of Workers,
DESIRING to deepen the solidarity between their peoples while respecting their history, their culture and their traditions ...
DETERMINED to promote economic and social progress for their peoples, taking into account the principle of sustainable development and within the context of the accomplishment of the internal market and of reinforced cohesion and environmental protection, and to implement policies ensuring that advances in economic integration are accompanied by parallel progress in other fields.[36]

Art. **2** proclaims the values of the Union,

The Union is founded on the values of respect for human dignity, freedom, democracy, equality, the rule of law and respect for human rights, including the rights of persons belonging to minorities. These values are common to the Member States in a society in which pluralism, non-discrimination, tolerance, justice, solidarity and equality between women and men prevail,

Article 3 contains the objectives:

1. The Union's aim is to promote peace, its values and the well-being of its peoples.
2. The Union shall offer its citizens an area of freedom, security and justice without internal frontiers, and an internal market where competition is free and undistorted.
3. The Union shall establish an internal market. It shall work for the sustainable development of Europe based on balanced economic growth and price stability, a highly competitive social market economy, aiming at full employment and social progress, and a high level of protection and improvement of the quality of the environment. It shall promote scientific and technological advance.

   It shall combat social exclusion and discrimination, and shall promote social justice and protection, equality between women and men, solidarity between generations and protection of the rights of the child. It shall promote economic, social and territorial cohesion, and solidarity among member states. It shall respect its rich cultural and linguistic diversity, and shall ensure that Europe's cultural heritage is safeguarded and enhanced.
4. In its relations with the wider world, the Union shall uphold and promote its values and interests. It shall contribute to peace, security, the sustainable development of the Earth, solidarity and mutual respect among peoples, free and

---

[36]Cfr. 30.3.2010. *Official Journal of the European Union*. C83.

fair trade, eradication of poverty and the protection of human rights, in particular the rights of the child, as well as to the strict observance and the development of international law, including respect for the principles of the United Nations Charter.

5. The Union shall pursue its objectives by appropriate means commensurate with the competences which are conferred upon it in the Constitution.

In art. 6.1, The Union recognises the rights, freedoms and principles set out in the Charter of Fundamental Rights of the European Union of 7 December 2000, as adapted at Strasbourg, on 12 December 2007, which shall have the same legal values as the Treaties. The rights, freedoms and principles in the Charter shall be interpreted in accordance with the general provisions in Title VII of the Charter governing its interpretation and application and with due regard to the explanations referred to in the Charter, that set out the sources of those provisions. In accordance with article 6.3 "shall constitute general principles of the Union's law".

From these premises, the European Union proclaims in its Title I "Dignity", (art. II-61): Human dignity is inviolable. It must be respected and protected; dignity includes (art. II-63) the right to the physical and psychic integrity of the person, prohibits eugenic practices, the selection of persons and the cloning of embryos, prohibits (art. II-64) torture and degrading human punishment or treatments, such as (art. II-65) slavery and forced labour. Title II "Freedom" groups together (arts. II-66 to II-79) the *negative* freedoms (from) and *positive* freedoms (to) and Title III "Equality" (arts. II-80 to II-86) articulates the classic equality *before* the law and the modern equality *in* the law, with special emphasis on the rights of children, women, handicapped persons, and minorities.

The Charter is a document which unites the past, present and future. It combines rules with values and principles. It joins the traditional subjective, negative, reactionary rights and rights of resistance to power with the modern fundamental rights that demand positive actions. It institutes new rights in relation to the integrity and the advances of bioethics. It accepts the "cultural, religious and humanist" heritage of the founding fathers that constitutes the foundation and the reason for universal values and the inviolable and inalienable rights of the person and proclaims "freedom, democracy, equality and the rule of law". In harmony with current constitutionalism it reaffirms the principles regarding rules and protects rights, even in the presence of a majority action (constitutionalism); it reinforces guarantees and limits power, submitting the states to the jurisdiction of the ECHR; it gathers into a commendable synthesis the axiological objectivism the methodological intersubjectivism; the pluralism that is different from the *polytheism of values* (Weber) and from relativism it impedes the imposition of a single and excluding order of values, guarantees that the Constitutions be those of *jurists and not of priests* (Denninger) and promotes respect for diversity and difference; lastly, the Charter sets out the limits of that which is intolerable that constitutes the rights (universal, inalienable and inviolable) of the person.

Its greatest accomplishment is leaving everyone unsatisfied, because no one can consider themselves the owner of rights. It confirms the personalist imprint that emanates from the respect for dignity and for the rights of the person; it recognises, in accordance with Liberalism, the rights of the individual and the classical freedoms and affirms social rights in line with Socialists. The proclamation of the rights inspired by solidarity and the attention to bioethics show there is an move towards the construction of a supra-statal and supranational *novus ordo, or new order*; however there are criticisms because the Charter has not regulated the new rights (family, homosexuals, gender) or because it has not incorporated the (Christian) roots of Europe.[37]

The Charter reaffirms the commitment of the EU to the social rights enunciated in the SCE (Turin 1961) and in the *Communitarian Charter of the Fundamental Social Rights of the Workers of 1989.* For the first time, the Title IV "Solidarity" (arts. II-87 to II-98) recognizes the rights of the *third generation* supported by the solidarity which breaks with the commutative reciprocity and synallagmaticism of the exchange of goods and services based on *do tu des* and demands the proportional equality inherent in distributive justice that respects the capabilities and concentrates on satisfying basic and radical needs. It is very significant that the subject of the rights detailed in this Title is "all", "every person", and not only citizens, even when there are referral clauses to the laws of the member states that limit them to citizens. Special attention is merited (as a counterpoint to deregulation and the current fragility of rights) for the emphasis on the protection of the rights of workers (arts. II-89 and 91), on the establishment of just equitable working conditions; on the prohibition of child labour and on the protection of young people in the workplace (so that cases like Nike or Reebok do not occur); on the attention (art. II-93) given to family and professional life, social security and social assistance, on the explicit commitment to combating social exclusion and poverty (art. II-94), and the protection of health (art. II-95) and consumers (art. II-98).

The *Joint Report on Social Protection and Social Inclusion 2009*, accentuates the importance of solidarity that supports its social policies withstand the impact of the economic crisis and favour recovery, based on the systems of social protection that constitute the nucleus of *Welfare.* The Commission recommends utilizing the potential of the European Social Fund to mitigate the effects of the crisis; adopting exhaustive strategies of active implementation directed towards inclusive labour markets, access to quality services and adequate minimum rents; supporting the states' fight against poverty and the social exclusion of children and child care; the lack of housing generates exclusion and marginalization that accentuates the vulnerability of immigrants, which in the case of Roma (gypsy) citizens is extreme; it warns about the new at-risk groups, e.g., youths and workers who have recently entered the working world; without an employment rate of 50% of older workers, the pension system is in danger; women and workers with low incomes are at high risk. The member states must improve health assistance and reduce inequalities focusing on primary care, prevention, health promotion, better coordination and the rational use of resources. The MAC (*Open Social Market of Coordination*) culminates in

---

[37]Cfr. VIOLA, F., Diritti e Identità, en VV. AA., *Identità e diritti in Europa*, a cura di. Valentina Strazzeri (Padova: Messagero, 2004), p. 56 y ss.

2010, the objective of the Lisbon strategy and the European Year for Combating Poverty and Social Exclusion.

In the European Year for Combating Poverty and Social Exclusion, the European 2020 strategy for smart, sustainable and inclusive growth puts forward three mutually reinforcing priorities, promoting employment and economic growth through:

(a) Three priorities: **Smart growth**: developing an economy based on knowledge and innovation. **Sustainable growth**: promoting a more resource efficient, greener and more competitive economy. **Inclusive growth**: fostering a high-employment economy delivering social and territorial cohesion.
(b) Five headline targets: 75% of the population aged 20–64 should be **employed**. 3% of the EU's GDP should be invested in R&D. The "**20/20/20**" **climate/energy targets** should be met (including an increase to 30% of emissions reduction if the conditions are right). The **share of early school** leavers should be under 10% and at least 40% of the younger generation should have a tertiary degree. 20 million less people should be at risk of poverty), and
(c) **Seven** flagship initiatives to catalyse progress under each priority theme: "**Innovation Union**" to improve framework conditions and access to finance for research and innovation so as to ensure that innovative ideas can be turned into products and services that create growth and jobs. "**Youth on the move**" to enhance the performance of education systems and to facilitate the entry of young people to the labour market. "**A digital agenda for Europe**" to speed up the roll-out of high-speed internet and reap the benefits of a digital single market for households and firms. "**Resource efficient Europe**" to help decouple economic growth from the use of resources, support the shift towards a low carbon economy, increase the use of renewable energy sources, modernise our transport sector and promote energy efficiency. "**An industrial policy for the globalisation era**" to improve the business environment, notably for SMEs, and to support the development of a strong and sustainable industrial base able to compete globally. "**An agenda for new skills and jobs**" to modernise labour markets and empower people by developing their skills throughout the lifecycle with a view to increase labour participation and better match labour supply and demand, including through labour mobility. "**European platform against poverty**" to ensure social and territorial cohesion such that the benefits of growth and jobs are widely shared and people experiencing poverty and social exclusion are enabled to live in dignity and take an active part in society.

8.8 The Civil Society Versus Poverty and Exclusion

Even in 2010, declared the International Year for the Eradication of Poverty and Social Exclusion these principles are, like the *Millenium Objectives* a theme still pending. From the *civil society* and the NPO's, *Zero Poverty*, impelled by Caritas Europa and based on the principles of respect for the person, solidarity, subsidiarity and the *common good* proclaimed by the Encyclical *Caritas in Veritate*, presents ten reasons to act and affirms that even today, social exclusion and poverty are living and emerging realities in Europe.

Poverty in the EU: Eurostat: 84 million people in Europe lack the two primary goods. (**2**) **Misery in Eastern Europe**: World Bank: 60 million persons in Eastern Europe and Central Asia live on $2 (1.5€) a day. **Illiteracy**: Unesco: 7.2 million people of Central Eastern Europe are illiterate. (**4**) **People without work:** Eurostat: 1 of every 10 persons live in Europe in a family where no one works. (5) **Social exclusion:** Eurostat: 8 out of every 10 Romanian youths suffer discrimination at school in Europe. (**6**) **Poverty among workers**: Eurostat: 1 out of every 10 workers lives in conditions of poverty. (**7**) **Poverty among the elderly:** Eurostat: 28% of those over 65 years of age live in conditions of poverty. (**9**) **Scholastic failure:** Eurostat: 1 out of 7 youths between 18 and 24 years of age does not finish his studies, and (**10**) **Absolute Poverty in the EU:** Eurostat: 17% of the European population is at risk of poverty.

8.9 Globalization and the Crisis of the Social State

The *subprime* crisis (July–August 2007), the Wall Street crash and the financial crash (Fannie Mae, Freddi Mac, Lehman Brothers, AIG) unleashed (August 2008) the economic crisis that drags the crisis of the social state with it and proves that globalization increases the wealth of speculators and impoverishes the workers. The world, Sen writes, is spectacularly rich and painfully poor.

The *fiscal crisis of the state* (O'Connor), the voraciousness of the *insatiable rights* (Pintore), the *unfulfilled promises* and the *unsatisfied demands* (Bobbio), the opaqueness, and the lack of ethical criteria to guide the actions of public and private institutions; the multiplication of social spending, the emphasis on rights and the silence regarding the corresponding duties and responsibilities and the Neoliberal challenge that promises greater efficiency, and does no justice to a *minimum state*, erode legitimacy, sow fear and propagate the insecurity that unleash crisis in the social state, whose continued existence, even in Europe, is at grave risk.

Globalization shows that the crisis of the national state (Westfalia) that no longer exercises its sovereignty absolutely and unlimitedly over the territory nor over its citizens has lost its fiscal empire. The monetary, social, fiscal, defence, justice and energy policies depend on a *supra-statal* entity such as the European Union. The guarantee of rights also depends on a *supra-statal* court, the EU's Court of Human Rights, whose decisions bind the member states.

The (failure of) multiculturalism and the handling of diversity; the integration of immigrants, international cooperation and respect for the rights and duties of all, show the problems facing the current ethical, political and social pluralism that accentuates the crisis of the state incapable of reaching even the *minimums of justice* that establish the limit of political legitimacy and of moral decency.

The fallacious (positive) impact of globalization and of the universality of rights has generated an expansive centrifugal force that overflows the social state. *Ad intra*, because the national state, weak with the strong and strong with the weak, is incapable of guaranteeing rights and of demanding fulfilment of (positive) duties, of redistributing fairly and equitably the primary goods, of attending to the capabilities and respecting its subjects, persons, active citizens, not *servant citizens*, who are neither *clients* nor *users*; because deregulation contradicts the *ratio* of the social state, which no longer generates equality or solidarity and multiplies spending; because corruption corrodes the legitimacy of power and because the Neoliberal strategy and the supporters of the *minimal sate* question its efficiency and efficacy. *Ad extra*, because the *positive* rights and duties demand planetary solidarity that collides with borders. Habermas, Bobbio and Held remind us that globalization limits the sovereignty of the national state and calls for a Kantian worldwide republic (*Weltrepublick*) of free states.

8.10 Conclusion

1. The founding fathers, committed to peace, social justice and economic development, convinced of the universality, inviolability and inalienability of human rights, understood the construction of the a free and socially supportive EU based on respect for the rights and the enforceability of duties.
2. The European Union has not been, nor is, nor will be (only) a commercial, customs, nor monetary union. From its constitution until now, its foundations, its identity, its values and its objectives pass through the establishment of peace, human development, social justice and equity starting from dignity, liberty, equality and solidarity. Dignity, freedom, equality and solidarity are the values that inspired the *founding fathers of the European Union* (Adenaeur, de Gasperi, Monet, Schuman) and that preside over the charter of the fundamental rights of the EU.
3. Dignity constitutes every human being; because of this, each person (*substantia individua rationalis naturae*), each one, *es res sacra est*; for this reason, *man* (each man) possesses *dignity and not price, and, he is "extra commercium"*; for this reason, when the dignity of a human being is violated, humanity suffers. Dignity constitutes the new *regula aurea*. For Carillo Salcedo, the recognition of the intrinsic dignity of every human being constitutes a radical revolution because in contrast with classic international law, the person can no longer be considered a mere object.[38]

[38]Cfr. CARRILLO SALCEDO, J. A., *Dignidad frente a barbarie. La Declaración Universal de Derechos Humanos cincuenta años después* (Madrid: Trotta, 1999), p. 16 y ss. Cfr. BLÁZQUEZ, J., *Igualdad, dignidad y libertad.* (Universidad Pública de Navarra, Pamplona 2005, 4ª ed. corregida y aumentada), p.13 y ss.

4. The recognition of and respect for dignity, from whence arise *positive* rights and duties, defines the current Constitution. Rightly did Truyol write that "the fundamental human rights have their root in the dignity and value of the human person".[39] UDHR (1948) "recognition of the inherent dignity and of the equal and inalienable rights of all members of the human family is the foundation of freedom, justice and peace in the world", as art. 1 affirms: "All human beings are born free and equal in dignity and rights".

   The present constitutionalism, replete with values and principles reunites the natural law tradition that sustains the principle of the dignity of the person rooted in his ethical autonomy as a moral being, with the positivist tradition that links the rights that are inherent to him with respect for the law. The *Charter of Fundamental Rights* of the *European Union*, inspired by the humanist ethics that arose from Christian Personalism, Democratic Socialism, and liberal philosophy, explicitly present in the *founding fathers* of Europe and in the post-war constitutions, and that even today, implicitly inspire the Treaties of the Union, proclaim *dignity, freedom, equality, solidarity, citizenship and justice* and constitute the best guarantee for human rights today.
5. The *social state* is an undeniable achievement of the European Union that has freed European citizens from fear, misery and misfortune and has configured the sense of justice in a *well-ordered society* (Rawls). Welfare is, above all, a state of *justice* that defines the threshold of indecency, well-being and affirms that human development that is not (only) economic growth. Since it was first published in 1990, *The Human Development Report 2010* confirms the initial perception: "People are the real wealth of a nation." (A. Sen) and shows that (human) development does not coincide with income.

   The social state of law, of *positive* rights and duties, strong, limited (by the law), and subsidiary; held in the public's mind and by *deliberative and participative* democracy,[40] that supports the SME and the *civil economy*, that oversees the economy and cares for civil society, is, unlike all the Neoliberals, the Anarcho-Capitalsits and the partisans of the *minimal state*; unlike those who still support the primacy of the *homo oeconomicus*, of benefit and of unlimited growth and unlike those who separate public ethic from private ethic, the (only) valid and correct alternative to come out of the present crisis.
6. Our time is the *time of rights* (Bobbio), of *positive duties* (Zagrebelsky) and of responsibility (Jonas). Before the multiplication and the extra limitation of absolute, unlimited and *insatiable* rights,[41] Cotta insists on the corresponding

---

[39]Cfr. TRUYOL SERRA, A., *Los derechos humanos* (Madrid: Tecnos, 1974), p. 29.

[40]Cfr. Sen, A, *The idea of justice* (London: Allen Lane; New York: Penguin Books, 2009). RAWLS, J., *Political liberalism* (New York: Columbia University Press, 1993).

[41]Cfr. PINTORE, A., "Derechos Insaciables", en FERRAJOLI, L., et alii *Los fundamentos de los derechos fundamentales*, ed. de A. Cabo y G. Pissarello, pp. 243–267 (Madrid: Trotta, 2001).

rights and duties, on the equality and symmetry of the subjects, on the reciprocity and proportionality of their actions and on impartiality in judgement.

7. The *common good* tempers the individualistic and utilitarian excesses, impels the theories of *global justice* (Rawls, Sen, Pogge, Trujillo[42]) and promotes the construction of a *worldwide, planetary and global ethic* (Kung). Respect for the person, the recognition of the other, the subsidiarity of the state, the civil economy, gratuity, reciprocity and solidarity define the reconstruction of the social state and fortify the *European social model.*
8. Person, (civil) society, (social) state. Jacques Maritain, (*Man and the State*, 1951), whose influence on the post-war constitutions and on the European Union is well known, proclaimed the order that contains the three principles of the social state: the primacy of the person, the priority of the *civil society* and the subsidiarity of the state.[43]
9. The Declarations of Rights, of *man and of the citizen,* remind us that the person precedes the citizen. Because one is a person, one acquires the condition of citizen and not *vice versa*. The recognition of the person, the respect for his *inviolable* dignity (Die würde des menschen ist *unantastbar*. Sie zu achten und zu schützen ist Verpflichtung aller staatlichen Gewalt. -art. 1.1. Grungesetz- : – human dignity shall be inviolable. To respect and protect it shall be the duty of all state authority) and the protection of rights, *inherent, universal, inalienable, and undeniable* (art. 2C. It. "La Repubblica riconosce e garantisce i diritti inviolabili dell'uomo, sia come singolo sia nelle formazioni sociali ove si svolge la sua personalità, e richiede l'adempimento dei doveri inderogabili di solidarietà politica, economica e sociale",[44]) which are "el fundamento del orden político y de la paz social" – the foundation of the political order and of social peace (art. 10.1C.Esp.) constitute the nucleus of constitutionalism and of the social state of law and show the moral quality of democracy founded on recognition, respect, protection and promotion of the rights and the enforceability of the *positive duties* oriented towards the *common good.*
10. Sen shows that the elimination of poverty strengthens the democracy that is made up of human security, *civil society, human* development (which is not only economic growth), and quality of life. The fight against poverty and exclusion demands the elimination of the obstacles that impede liberty, such as

[42]Cfr. POGGE, T., *Kant, Rawls and global justice* (Ann Arbor: UMI Dissertation Services, 1983). RALWS, J., *The Law of Peoples: With the Idea of Public Reason Revisited* (Baltimore: The Johns Hopkins University Press, 1999). TRUJILLO, I., *Giustizia globale. Le nuove frontiere dell ´eguaglianza* (Bologna: Il Mulino, 2007).

[43]Cfr. BEA, E, i COLOMER, A., "Mounier i Esprit: Europa contra les hegemonies" en Quaderns Uniò de Traballadors Democratacristians de Catalunya (4 Barcelona, 2001, págs. 11–73. COLOMER, A., MONZÓN, A (eds) Enmanuel Mounier i la tradició personalista, (València, PUV, 2002).

[44]"The Republic recognizes and guarantees the inviolable rights of man, both as a single person as well as in the social formations where he develops his personality, and requires the fulfillment of non-transferrable duties of political, economic and social solidarity" [Trans.].

hunger and misery, tyranny, illiteracy, inattention to social and sanitary issues, and a lack of care for the environment, the lack of freedom of expression, the intolerance of others that limits the *public use of reason* and the real possibility to choose in an autonomous and free manner the life that each person judges best. This requires a strong, limited and subsidiary (federal) state and a free, social and supportive civil society. This is the great goal that the European Union must face today to fulfil the *Millenium Objectives,* reduce poverty and achieve human development, social well-being, justice and peace.

Chapter 9
Peace as a Priority

Pedro Talavera

Abstract Chapter 9 underlines how since its inception the European Union has presented itself to the world as a great "peace project". The European project consists in creating a "community of peace and democracy" based on the conviction that peace does not emerge from dissuasion but rather out of cooperation. Europe attempts to remodel the world by realising the Kantian dream of "perpetual peace", which is not based on the balance of military powers (or the supremacy of one of them) but rather on the universal recognition of the dignity of the human being, on the universal guarantee of human rights, and on sustainable human development, which means superseding the current neoliberal model of globalization. From this perspective, the chapter analyses the confrontation between the Hobbesian paradigm of "the inevitability of war" and the first theoretical models which tried to establish the possibility of lasting peace in Europe. Secondly, the discussion also considers the failure of the bellicist paradigm and Schmitt's theses, which are taken up by North American 'neocon' ideology and responsible, to a large extent, for its current decadence in relation to the emergent European model, which reformulates the Kantian paradigm as realisable, and which has become a global point of reference.

9.1 Introduction. Europe as Peace Project

In the deepest of European thinking, Europe has always been considered a "peace project". Immediately after the Second World War, the founders of what would become the European Union wanted, above all, to prevent another war on European territory. From their modest beginnings with ECSC, to the Treaty of Rome and the Economic Community, the "European project" has always been driven by the

P. Talavera (✉)
Philosophy of Law, University of Valencia, Valencia, Spain
e-mail: Pedro.Talavera@uv.es

J. Ballesteros et al. (eds.), *Globalization and Human Rights*, Ius Gentium: Comparative Perspectives on Law and Justice 13, DOI 10.1007/978-94-007-4020-4_9,

pursuit of peace. This is made irrefutably explicit in the *Schuman Declaration* of 1952: "World peace cannot be safeguarded without the making of creative efforts proportionate to the dangers which threaten it (. . .). Europe will not be made all at once, or according to a single plan. It will be built through concrete achievements which first create a de facto solidarity. The coming together of the nations of Europe requires the elimination of the age-old opposition between France and Germany (. . .). The pooling of coal and steel production should immediately provide for the setting up of common foundations for economic development as a first step in the federation of Europe (. . .). The solidarity in production thus established will make it plain that any war between France and Germany becomes not merely unthinkable, but materially impossible".[1]

Peace as a priority has been the axiom upon which all activities directed towards building a united Europe have been based. In fact, after the end of the Cold War, the European Economic Community was conceived as a fundamental instrument for overcoming the East-west tension in Europe. And the actual (economic and political) project of constructing the European Union has far exceeded the *Westphalian* model, based on a politics of balance and on the defence of nation-state sovereignty, out of the conviction that peace does not emerge from *dissuasion*, but rather out of *cooperation*; it is not born out of *imposition* but instead from the *perception* of real expectations of development. This thinking has enabled Europe to overcome the limited paradigm of "fortitude", in order to become an open, plural, stable and solid community in which the secular concerns of its citizens with regards to security have gradually diminished until being almost forgotten, and in which "preventing war" and "building peace" appear as powerful motives in their minds (the *2009 Eurobarometer*, indicates that for 89% of those surveyed, "maintaining peace and security" is a priority of the EU and the most effective of its policies).[2] In today's Europe, it is not just that war is undesirable, but rather that it is absolutely inconceivable.

Europe presents itself as a political and economical entity that, in only 50 years, has eradicated any idea of war among its powers, whose economic potential has come to equal that of North America, and whose external strategy has managed to free a number of countries from dictatorship and to set up a reliable democracy within them. Furthermore, the influence of this "peace zone", which takes in 480 million citizens, extends to another zone of 385 million people who share land or sea borders with the EU, and to another 900 million who live in areas which enjoy preferential treatment with the EU with respect to commercial, investment, credit

[1]Shuman Declaration: http://ec.europa.eu/publications/booklets/eu_documentation/04. See P. FONTAINE's analysis, *A New Idea for Europe. The Schuman Declaration (1950–2000)*, 2nd ed. (Luxembourg: European Documentation, 2000).

[2]MØLLER, Bjørn., "*National, Societal and Human Security. A General discussion with a Case Study from the Balkans*". Paper for the First International Meeting of Directors of Peace Research and Training Institutions on What Agenda For Human Security in the Twenty-first Century? (París: UNESCO, 2000) [Quoted from: B. Moller, "Seguridad nacional, societal y humana: el marco general y el caso de los Balkanes", *FASOC*, año 15, n. 4 (octubre-diciembre 2000), pp. 9–10].

or humanitarian ties. In other words, the EU has a zone of influence known as the *Eurosphere* that covers almost 2,000 million people, and which is gradually being transformed by the *European Project* and adopting *European way* of doing things.[3]

This *European project* consists specifically in creating a "community of peace and democracy" which is becoming increasingly wider. With this in mind, the *European way* of doing things is not an imposition (the force of military power), but rather a "seduction"; the attraction of peace and democracy through its market strength and the expectations linked to it. In this way it achieves the most important of all objectives: the remodelling of societies from the inside. No doubt some analysts (particularly North American ones) identify this *European way* of doing things with "weakness" and with a lack of effective solutions in the face of the serious security problems that threaten Western societies. But the results, however, support the power of this "weakness".[4] The *European project*'s vocation is one of transforming and therefore it functions in the long term (it does not attempt to win battles in the short term), because it attempts to remodel the world by establishing the necessary premises for realising the Kantian dream of "perpetual peace", which is not based on the balance of military powers (or the supremacy of one of them), but rather on the universal recognition of the dignity of the human being, on the universal guarantee of human rights, and on sustainable human development which entails superseding the current neoliberal model of globalization.

In actual fact, the *European project* is cemented upon the *European dream*; upon an ideal of personal transformation that exceeds the strict horizon of material well-being and the accumulation of wealth, and aims at an elevation of the human spirit based on the centrality of the person, on the recognition of their inalienable dignity and on the guarantee of universal human rights, as stated in the EU's Charter of Fundamental Rights. This is a dream that, just as Rifkin has shown, emerges at the same time that the American dream, having aged and become rather old-fashioned, begins to wane because it is no longer connected to the aspirations of a plural and global world. Its conception of success has been too tightly linked to financial success; too centred on autonomy, self-sufficiency and material progress and too little concerned with the progress of the global human society to seduce an interdependent and multicultural world. The European dream, by contrast, places more emphasis on community relations than on individual autonomy; on cultural diversity rather than assimilation, on quality of life instead of the accumulation of wealth, on sustainable development more than unlimited economic growth, on human rights over and above property rights, and on global cooperation more than the unilateral exercise of power. What the European dream aims for, in summary, is the creation of a new history.[5] It is a powerful dream precisely because it

---

[3]LEONARD, M., *Why Europe Will Run the 21st Century* (London: Fourth State, 2005). [Quoted from: *Por qué Europa liderará el siglo XXI* (Madrid: Taurus, 2005), p. 15.]

[4]*Ibíd.*, p. 16.

[5]RIFKIN, J., *The European Dream: How Europe's Vision of the Future Is Quietly Eclipsing the American Dream* (New York: Jeremy P. Tarcher/Putnam, 2004) [Quoted from: *El sueño europeo.*

dares to suggest a new course for history which emphasises the central and universal value of human dignity, sustainability, peace, human rights, and quality of life as a consequence of these values.

This dream now forms part of the current imaginary of Europe as a community, and of Europeans as citizens of the EU, and is gradually spreading across its political and economic sphere of influence. Europe has been referred to as an "invisible hand" that operates by way of traditional political structures.[6] And that is because the European project has managed to penetrate Europeans' lives, infiltrating the core of existing structures in the national life of the States; leaving institutions apparently intact but transformed internally. This was one of Jean Monnet's greatest insights: not to try to abolish the nation-state (nor nationalism), but rather to change its character, moving towards a less exclusive concept of sovereignty. And that is what turned Europe into a powerful global player. Each Member State is at one and the same time an agent of the EU. Community policies and values are put into practice by way of States' institutions, thereby enabling Europe to extend its influence without becoming the object of hostilities. While every American company, embassy or military base is a terrorist target, the relative invisibility of Europe allows it to expand its global network without this being interpreted as provocation.[7]

Its influence is not found, however, in the classic game of geopolitical strategies, but rather in the fact that Europe is "obsessed" with an insistence on legal frameworks. The European model for transforming societies and establishing peace and democracy comes about through Law. The Unites States may have changed the regime in Afghanistan, but it has not changed the society and therein lies its great failure. Europe has changed the societies of its 12 most recent incorporations and it has done so through its legislations, from health to agricultural, or from work to property, and taking in minority rights as well as freedom of religion. Moreover, it is also managing to change the societies of some countries who aspire to join the club, such as Turkey, overcoming and integrating broad sources of internal resistance in a peaceful way.[8]

On the other hand, by creating the world's largest single internal market, the EU has become an economic giant; but what is essential about this does not lie in its

---

Cómo la visión europea del futuro está eclipsando el sueño americano (Barcelona: Paidós, 2004), pp. 13–18].

[6]LEONARD, M., *Why Europe Will Run the 21st Century*, p. 29.

[7]Europe becomes invisible even in the strictest military sphere. For example, when European troops go abroad on international missions they very rarely use EU uniforms. On most occasions they serve under UN or NATO flags. And in the protectorates entrusted to them (Bosnia or Kosovo), their representatives act on behalf of both the UN and NATO. Europe's foreign policy strategy of invisibility could partly lie in its dark colonial past, which would help to explain why it is reticent to adopt imperialistic attitudes. There is, however, a deeper reason; for Europe, difference and plurality are not a threat, which also helps to understand its respect for cultures and identities and its aim to promote coexistence and peace amongst them.

[8]LEONARD, M., *Why Europe Will Run the 21st Century*, pp. 67–69.

size but in the nature of its economic model which is a synthesis of the freedom and initiative of liberalism, together with the social protection, stability and well-being characteristic of social democracy. A synthesis, that is, which is showing to be highly favourable and effective in terms of generating conditions for peace and development.

Having said that, it does not mean that Europe does not have serious defects and cannot be accused, for example, of meanness in its immigration policies, irrationality in its agricultural policies, abuse in its legislative production, hesitancy in its external politics, and so on.[9] All these aspects can support the accusation that Europe has still not reached the height of its dream (of its great peace project). Nevertheless, its project is solid and the priority of peace within it is unquestionable: so much so that this century has already been entitled the "new European century".[10]

With its contradictions, limitations and challenges, Europe is, then, an extraordinary and ambitious project for peace and for the construction of peace. But what is the European conception of peace? How has it been formed and upon which anthropological foundations is it based? How does it differ from the North American conception? We will attempt to unravel these questions in what follows.

9.2 Europe and the Dream of "Perpetual Peace"

Today, European countries wholeheartedly refuse the kind of *power* politics which has ruled during centuries and which has led to so much death and destruction in the world. After centuries of fighting each other, the 27 European nations have now put down their weapons, united, and forged an alliance which has pushed the spectre of war away from them. This is due to a change of direction in their conceptions of state, power and foreign policy. For the last five centuries (and especially from 1945), the central concept of European politics has revolved around the *balance of power* between their leading nations in order to deter the hegemonic temptations that emerged from the Peace of Westphalia in 1648.

---

[9]Habermas is not as optimistic and, given the inexistence of a Europeran identity, he thinks of European integration as problematic. In his view, the European project has three latent problems: the challenges of extending towards the East (unbalanced political and electoral weight); the political consequences of economic integration, especially in those countries belonging to the Euro-zone (unbalanced distribution of costs and benefits); and the changes facing the world political situation (the requirement of a solid European position regarding security and defence). See HABERMAS, J., *The Divided West* (Cambridge, UK: Polity Press Limited, 2006). [Quoted from: *El occidente escindido* (Madrid: Trotta, 2006), pp. 72–78]. Nevertheless, by accepting the need to overcome the problems inherent to the process of European unification, especially transcending bureaucratisation and consolidating a transnational conscience in the citizens which is not linked to nation but bound to the Constitution (*Ibíd.*, p. 84); in our view, that which is decisive is the latent persuassion (in elites and citizens) of the possibility of achieving it, as well as of the fact that peace, the dignity of the human being, and human rights are the indisputable premise of that integration. And this is an exciting project that inspires Europeans to continue its realization.

[10]LEONARD, M, *Why Europe Will Run the 21st Century*, p. 19.

The doctrine of *balance* dates back to Lorenzo de' Medici, whose family ruled fifteenth century Florence. During that period, Florence was one of the five city states that dominated the Italian peninsula alongside Rome, Naples, Venice and Milan, with whom it shared immense wealth and against whom it fought interminable struggles. In 1454, Francesco Sforza, Governor of Milan, approached the Medicis to propose an alliance between both states that would help them defeat Venice in order to gain it as an ally and thus halt the expansion of Rome in the future. In his response, Lorenzo de' Medici pronounced the historical sentence: "Italian affairs must be kept in balance". And such is the principle that has been in effect for 500 years: a system based on the mechanical idea of the scales according to which the different states had to outweigh each other so that neither of them could dominate the continent.[11]

This old ideology became the incentive that helped the European leading nations to develop the world's most advanced technology which allowed them to overcome the Eastern Empires and take control of the world. Nevertheless, a conception impregnated with the Hobbesian maxim of "war of all against all" also carried within it the terrible reality of '*perpetual war*': the aim of all subsequent wars, including the Thirty Years War, the Franco-German wars, the First and Second World Wars and the Cold War was to avoid the hegemony of another country and were founded on Hobbesian pessimism-realism.

This outlook is now history in Europe. Today, nobody fears the advance of Germany or France because Europeans have formed a *network* which, while respecting their differences, unites them under the blanket of Economy and Law.[12] The doctrine of balance has been substituted once and for all with a new perspective based on building *perpetual peace*. This new European Project is rooted in the reformulation of an old dream, to which the well-known essay *Perpetual Peace: A Philosophical Sketch*, published by Inmanuel Kant in 1795, aspired. Although the essay barely attracted any attention at the time of its publication, it was retrieved some years after the Second World War and has now become an essential reference point for theorists of pacifism as well as for most European intellectuals.[13]

---

[11]BOBBIT, P., *The shield of Achilles: War, Peace and the Course of History* (London: Penguin, 2003).

[12]There is a question that keeps recurring: if great powers have always provoked other countries to unite against them, why not Europe? The easiest answer consists in believing that Europe is only an economic – but not political - power (the economic is attractive whereas the political generates hostility). But that is not really the case, because Europe has become a network rather than a state. States' relations, according to the well-known paradigm, are like snooker balls; they have a hard layer on the outside and they collide and bounce back when touching each other. But it is difficult to collide against a net, because the net does not have the hard core that stops it from colliding yet, at the same time, allows it to adapt to diversity. See LEONARD, M., *Why Europe Will Run the 21st Century*, pp. 43–44.

[13]Some of the most notable would be: HABERMAS, J., "The Kantian Project and the Divided West" in *The Divided West* (Cambridge, UK: Polity Press Limited, 2006); FERRAJOLI, L., *Razones Jurídicas del pacifismo* (Madrid: Trotta, 2004); BECK, U., "War is Peace: On Postnational

The fact is that the utopia of the old Kantian ideal of perpetual peace has prevailed over the more modern Kelsenian project of *Peace Through Law* (1944), and has managed to embed itself at the centre of the new European dream of aspiring to a world-order based on inclusion and consensus.[14] Although both authors share the same aim, they differ in their ways of achieving it. The most profound limitation of the Kelsenian project of peace resides in the very core of its own positivist conception and in the reduction of Law to a mere instrument of power and coercion, whereby its original consideration as a guarantor of non-violence and universal respect for the other, according to the classical principle of *neminem laedere*, of not harming the other, becomes blurred.[15] While Law may prevent, suppress or punish violent actions, it will never be able to eliminate the intention to attack. And herein lies the need to recognise something other than the perfectioning of juridical instruments and a sanctioning body, and the need to investigate the true anthropological bases of peace.[16] Two aspects are fundamental to this investigation; the recovery of the notion of the ontological dignity of all human beings, and the recognition and guarantee of universal human rights. It has been argued that the European model evolves towards peace through a kind of synthesis between the Kantian and the Kelsenian ways; that is to say that it moves towards the combination of a universal ethical consensus (represented by the universal recognition of the dignity of the human being and human rights) and a juridical rationality that helps to perfect the instruments of international law as well as strengthening international institutions and bodies (particularly by using military force solely for peacekeeping missions and under the direction of NATO, fully supporting the International Criminal Court jurisdiction, and so on).[17]

On the other hand, the American model continues to be anchored in unilaterality and has not managed to free itself from the Rousseaunian paradigm of exclusivist patriotic exaltation, the Hobbesian premise of *perpetual war*, and the Schmittian dialectics of Friend-Enemy. This is a model that continues to revolve around its superior military capacity and its incontestable control in this domain, which allows it to act as a sort of "global police" that imposes discipline by means of intimidating (Iran, North Korea) and punishing transgressors for their evil actions (Afghanistan, Iraq). These are the two diverging cosmovisions represented today by Europe and the United States[18] which represent none other than a continuation of the intellectual battle between the two current streams of thought concerning peace: the pacifist stream, inspired by Stoicism, and the bellicist stream, inspired by the values of Republicanism. The two streams respond to two conflicting visions of the world,

---

War" in *The Cosmopolitan Vision* (Malden: Blackwells, 2006). In the North American context we could highlight: RAWLS, J., *The Law of Peoples* (New York: Basic Books, 1993).

[14]HABERMAS, J., *The Divided West*, pp. 113–116.

[15]BALLESTEROS, J., *Repensar la paz*, pp. 113–114.

[16]See, *Ibíd.*, chapter IV; pp. 108–131.

[17]RIFKIN, J., *The European Dream*, pp. 396–397.

[18]As seen by RIFKIN, J., *The European Dream*, pp. 377–381.

both founded on very different anthropological and historical postulates. Both streams have coexisted over the centuries, intertwining with each other and also opposing each other.

9.2.1 *The Idea of Perpetual Peace Versus the Idea of the Inevitability of War*

The idea of perpetual peace developed in Europe in parallel with another stream of thought, which was also European, but which praised war more or less explicitly. The struggle between both streams culminates in the eighteenth century in a radical divergence between Republicanism and Cosmopolitism, intellectually represented by the thought of Rousseau (which substantially matches Hobbes' regarding the inevitability of war) and that of Kant (who was convinced of the inevitable moral progress of humanity and its horizon of peace). The derivations of both positions have conditioned the idea and the establishment of peace right up to today.

The position which we could – perhaps abusively – call *pacifist* is rooted in the Stoic doctrine which, as is well known, emerges with the opening up of Greece to the world as a consequence of Alexander's conquests. Stoics shattered the exclusivist and quasi-xenophobic ideology of the *polis*, an ideology adapted to closed societies which were highly autarchic and withdrawn within themselves, and which nourished warlike values as an essential means of maintaining their independence from the surrounding empires. Since the notion of permanent army was still unknown, the *polis* was safeguarded by a militia of armed citizens. And here lies the importance that, for instance, Sparta gave to giving military training to its citizens, which constituted the main part of their education. It has been argued that this *polis*, the prototype of the city-state, was an immense military compound where, between the age of 20 and 60, the *equals*, as Spartan citizens called themselves, trained incessantly in the handling of weapons. Their life was dedicated to participating in public affairs and, as the axis of public life, to combat, which was the greatest means of social cohesion for the citizens of the *poleis*. But war was also the main resource for obtaining the slaves that fed their productive system. Likewise, it was also a way of relieving tensions when, at times of overpopulation, land was scarce and the great colonizing movement was therefore put into motion accompanied by subsequent wars. War thus played such a central role in these societies that a Spartan citizen, for instance, could lose their right to citizenship for lacking courage in combat.[19]

This way of life, that fosters patriotism and full submission to the community, suffers radical changes with Hellenism. The Greek world is transformed, the *poleis* sink with the institution of the monarchy, political life vanishes and values change. The Stoic doctrine spreads new ideals, such as universal brotherhood and equality amongst all human beings regardless of their socio-economic conditions,

[19]On this issue see the classic text: DE COULANGES, F., *La cité antique* (París: Hachette, 1927).

thus widening the restrictive concept of equality amongst equals that was so characteristic of the city-state. As a participant in universal *Logos*, man acquires a dignity unknown up to that point in the *polis*, whose disdain toward unequals and barbarians (those inferior beings who, in the words of Aristotle, must be conquered) was notorious. On the other hand, Stoicism preaches the cosmopolitan ideal in a world open to all human beings, where the right to citizenship may be shared by all in a universal *polis*. War would not make any sense in this society ruled by reason, which would render social institutions unnecessary. Nevertheless, and until the kingdom of reason came about, the stoic recommendation was to do good to your enemy (an instruction that Christianity would take to the point of offering the other cheek). As a peacemaking doctrine that promoted humanitarian feelings with slogans such as "comfort those who suffer" or "serve thy neighbour", Stoicism left its imprint on Roman customs and law, which experienced a shift towards equality. Rome became aware of the need to have a Law that was broad enough to include all human beings, not just its citizens, and passed laws that aimed to protect slaves from the cruelty of their owners, granted women legal capacity, and issued a "law of peoples", the *ius gentium,* common to all nations.[20]

The struggle between universalist and exclusivist values reaches modernity and is represented in the conflict between pessimist conceptions regarding the inevitability of war, incarnated by Hobbes and Rousseau and implicitly present in the juridicist peace proposal suggested by Kelsen, and the utopian-pacifist conceptions of Crucé and Saint-Pierre, assumed in Kant's ethico-political philosophy.

9.2.2 Hobbes and the Paradigm of 'Perpetual War'

From the sixteenth century onwards, with the worsening of armed conflicts resulting from the wars of conquest and religion, a new stream of thought emerges in Europe which aims to investigate the phenomenon of war and to put forward basic premises for peace. Western thought had traditionally submitted to the theory originating from the Bible of the perversity of human nature. With a few exceptions, such as Seneca in his *Moral Letters to Lucilius*, most thinkers assumed that war was inevitable given man's evilness. On that basis, the philosophers and moralists of the Renaissance could offer few alternatives beyond the moral condemnation of war and the exposition of its devastating effects, appealing to rulers to control their warrior impulses, and developing treatises to educate princes in meekness and other Christian values.

It would not be until the seventeenth century that, with the emergence of rationalism, a decisive inflection with regards to war would take place, transcending

---

[20]LONG, A. A., *Hellenistic Philosophy: Stoics, Epicureans, Sceptics* (Berkeley: University of California Press, 1986). [Quoted from: A. Long, *La filosofía helenística: estoicos, epicúreos, escépticos* (Madrid: Alianza, 1997), pp. 23–45].

the typical lamentations of moralists in order to produce political proposals and rational models of social organization that aimed to deactivate confrontation amongst human beings. These proposals were born in the wake of the discoveries made in physics and astronomy, having established that while nature is ruled by harmony and regularity, human society appears to be sunk in chaos and arbitrariness. This contradiction proved to be especially wounding to those who believed in reason and wanted to rationalize a world that they perceived to be ungovernable at all costs.[21]

No other thinker in this field knew better than Hobbes how to describe this realm of unsociable sociability (of which Kant would speak more than a century later) and which all authors of the period would attempt to resolve. But Hobbes' work also denotes a change in mentality. It no longer seeks to blame the human being, nor to incriminate it for its evilness like the moralists (from Montaigne to Pascal and La Bruyère), but rather eliminates all moral considerations stating that they should be understood as a combination of reason and passion. His texts refer to the overcoming of Cartesian dualism, but in exchange for an ambiguous conception of man; an ambiguity that will be taken up later by Spinoza who refuses to pronounce judgements about the goodness or evilness of a being that, as another part of nature, is determined by universal laws and, in particular, by the law of self-preservation.[22]

Hobbes is the great subversive, full of harshness and cynicism, who breaks with the Platonic and Aristotelian heritage and marks the dividing line that inaugurates modernity. It is he who brings about a change in the concept of justice and who defines it (in contrast to the moral notion of scholasticism) as the mere fulfilment of agreements. It is also he who, aseptically, separates the moral from the economy and negates the existence of a "fair" price (as if, he quips, it was unjust to sell something for more than one had paid for it).[23] And it is he who dares to recognise that the value of a human being is, as in the case of surplus goods, the price established by the market.[24]

Hobbes not only draws a sad psychological portrait of seventeenth century man within the framework of the development of capitalism, but also employs an analysis of social relations that will be taken on, in one way or another, by all subsequent authors: from Spinoza to Locke, from Montesquieu to Rousseau. Despite the criticism and reticism that his work aroused (prohibited in 1666 under the accusation of atheism and burned in Oxford after his death in 1679), his model would end up being an obligatory reference point and the model which has proved most influential in the internal politics of States up until today. Whether manifestly or latently, his most profound anthropological conception lies in the state of "war of

[21]VILLAVERDE, M. J., "Europa y el sueño de la paz perpetua", *Cuadernos de Alzate* 26 (2002): 27.

[22]BALLESTEROS, J., *Sobre el sentido del derecho* (Madrid: Tecnos, 2002), pp. 39–40.

[23]*Ibíd*, pp. 30–35.

[24]HOBBES, T., *Leviathan* (New York: Penguin, 2007). [Quotes from: T. Hobbes, *Leviatán* (Madrid: Editora Nacional, 1983), pp. 190–191].

all against all" that he describes: perpetual war is what defines the relations between men hungry to satisfy their appetite, and whose happiness consists in a constant race in pursuit of the attainment of one desire after another, a race that only ends with death. In the pursuit of those scarce goods coveted by all – riches, glory, power – desires and interests intersect and war breaks out: a war without end despite the fact that the voice of reason demands again and again that peace be sought in order to end an unsustainable situation. The demand for peace is made by way of the two primary laws of nature, the rational principles of coexistence and peace that man, dominated by passion, does not hear.

In his mechanistic vision, so characteristic of seventeenth Century Rationalism, Hobbes describes the laceration whereby man is split between conflicting forces. But the dominant vector, that of passion, is the one that imposes itself. Nothing could provide a reason for releasing the human being from this state of war if it was not for the cooperation of passions such as fear of death, or a sweeping passion that, allied with reason, induces them to make pacts. Through the social contract, men choose a political power by consensus, whose mission essentially consists in ending war. In this way, the State appears in *Leviathan* as the "lord of peace" to which much-prized freedom would be sacrificed in the name of security, and to which obedience would be surrendered as long as it accomplished its mission. To the reproach that the condition of the subjects in such a state would be fairly miserable, Hobbes responds unperturbed that man's condition could never be without some discomfort.[25]

In general terms, this is the alternative to the problem of war contemplated by the majority of authors in the seventeenth and eighteenth centuries; to establish a state that imposes peace. While it is true that significant variations exist with respect to the safeguarding of particular rights and with regards to the description of the nature of the state, in all cases the state is configured as the framework of rational legality in which individuals pursue their happiness in peace. However, it is obvious that such an institution would not end the problem of war and would represent merely a solution for achieving internal peace in the heart of the political community. The problem of war remains and is merely displaced to another area; that of the external sphere. Beyond its borders, each State remains in a situation of permanent conflict with other nations. To establish peace in this external context would be the challenge which some authors of the seventeenth century would face, and which would dominate the political scene of the eighteenth century.[26]

[25]HOBBES, T., *Leviathan.* p. 277.

[26]ZARKA, Y. C., *Hobbes et la penseé politique moderne* (Paris: PUF, 1995). [Quoted from: Y. C. Zarka, *Hobbes y el pensamiento político moderno* (Barcelona: Herder, 1997), p. 31].

9.2.3 The Peace Proposals of Crucé and Sully

In the wake of the Spanish theologians of the sixteenth century, such as Francisco de Vitoria and his *De iure belli*, seventeenth century thinkers like Grocio tried to establish norms in order to regulate relations between States and to impose rational order in international society. *De iure belli ac pacis* of 1625, which was published 11 times in Latin and 12 in Dutch, demonstrates the interest that the subject awoke during this period. With "peoples rights" – which would later come to be known as international law – having been secularised (Grocio insisted that natural right would exist even if God did not), a journey began which would reach its notorious peak in the eighteenth century, both in terms of doctrinal as well as institutional plans. The debate about the legality of war and the conditions that made it justifiable also heightened. Moreover, awareness surrounding the question of the conquest of indigenous peoples, which had been raised with recriminations and condemnation by the Spanish theologians from Brother Bartolomé de Las Casas to Father Vitoria, would continue to grow until culminating in the Kantian declaration against colonialism.

Within this gradual humanitarian consciousness-raising, the wars of religion of the sixteenth century had represented a milestone, leaving people and states exhausted and fostering a demand for peace that would extend across Europe. One of the highest indicators of this first pacifism that emerged in the seventeenth century was the book by Emeric Crucé, *Le nouveau cynée*, published in France in 1623.[27] Crucé, son of a member of the League, the extremist catholic party that had brought about the massacre of Huguenots during the St. Bartholomew's Day massacre of 1572, had been traumatised by the violent acts which he lived through during his childhood. His writings display a radical difference from another peace plan proposed at the same time by Maximilien de Béthune, Duke of Sully, contained in his memoirs which were published in 1640 and 1661 and known as the '*Grand Dessein*'.[28] While Crucé's book, which inspired it, fell into oblivion, Sully's project became an essential reference point for all subsequent peace treatises and projects such as that of the Abbé de Saint-Pierre. Nevertheless, Crucé's text is of greater importance, being a perfect demonstration of tolerance, humanitarianism and cosmopolitism, and an attempt to try to safeguard the interests of all nations and to respect their possessions. No other previous treatise, with the exception of Kant's, shows such an elevated conception and such an authentic love of humanity. In stark contrast to the usual rhetoric, his vision of the earth as a city common to all peoples, in which the right of hospitality (which Kant would later take up) reigns, comes close to the universalist conceptions of the stoics.

---

[27] CRUCÉ, E. *Le nouveau cynée ou discours des occasions et mohines d´eftablir une paix generale, & la liberté du comerse par tout le monde,* Paris, 1623.

[28] BARBICHE, B., *Sully. L'Homme et ses fidèles* (Paris: Fayard, 1997).

Sully's text proceeds along different lines. He is not motivated by humanitarian feelings such as the suffering and misery of people condemned to wars without end, nor to cosmopolitan ideals like those that drove Crucé, but rather by a pragmatic and political objective: to weaken the Habsburg dynasty. But to speak of peace within the framework of the Thirty Years War[29] is to play down the significance of a plan that would completely upset the European map, change borders and cut States into pieces. Hiding behind the geopolitical imbalance of Europe which, from the sixteenth century, becomes a permanent source of tension and conflict in terms of international relations, it reduces the number of European States to 15 eliminating the Russian and Turkish Empires.[30] This "Europe of the fifteen" would be governed by a general council – an idea taken from Crucé – made up of 60 plenipotentiaries who would regulate relations between each sovereign and their subjects, and would have the authority to stop tyrannical governments from setting up within the parameters of the member states.[31] In this way, Europe would enjoy perpetual peace.

If Sully's proposal was limited to the European sphere, and appealed to a union of Christian Europe against the Turkish Empire,[32] recalling the medieval "*Civitas Cristiana*", Crucé's book is a declaration in favour of universal peace that transcends different religions and is committed to reaching an agreement between Christians, Turks, pagans and Jews. Within this federation, the representative of the Turkish Empire, and the kings of Persia, China, Mongolia and Tartary, would sit alongside the Christian monarchs and the Pope. In contrast to Sully's goal of establishing a European Alliance to wage war against Muslims, falling into the contradiction of wanting to establish peace by way of war, Crucé demonstrates tolerance and a respect for diversity of beliefs which is extraordinarily modern. He asks for respect for all religions arguing that, despite differences in their rites and ceremonies, they all pursue the same end. Why should we persecute those who claim miracles and offer evidence for the adoration of God? Piety is incompatible with antagonism, he affirms, and differences in religion should not obstruct world peace. Crucé does not urge Christian monarchs to unite against the Turks or encourage the weakening of some powers in order to favour others as Sully does. He uncovers the reasons that

---

[29]The Treaty of Westphalia, which put an end to the Thirty Years War, dates from 1648, 8 years after the publication of the first two volumes of his memoirs.

[30]Which would be composed by six hereditary kingdoms: France, Spain, Great Britain, Denmark, Sweden and Lombardy; six elective states: the Papacy, Venice, the Empire, Poland, Hungary and Bohemia; and three federative republics: the Helvetic Republic, the Republic of Italy and the Republic of Belgium.

[31]Six particular recommendations would resolve the differences betwen states. The General Council would establish the contingent of men that each power would have to supply as well as the corresponding economic contributions, and would manage the military operations and the share of conquests.

[32]Despite the existing privileged relations between France and the Ottoman Empire, from 1536 the date of the capitulations between Francesco the First and Soliman the Magnificent. Crucé, on the contrary, granted the Sultan a preeminent rank, placing him after the Pope and before the Emperor.

incite war and launches a strong attack against one of these – honour – which he describes as a miserable virtue when it is achieved by way of bloodshed. To kill, he adds, is easy given that even the stupidest of animals are capable of it. Warlike virtues such as bravery, courage, daring, heroism, and so on should be knocked down from the privileged spot they occupy on the scale of societal values and substituted with humanitarianism. In effect, no cause could justify war for Crucé, not even the regaining of a usurped throne. The greatness of a king should not be measured by the size of his kingdom, he says, but by his contribution to the propagation of world peace. Only this way would he ensure that his name appeared in golden letters in the chronicles of history. In an era which valued warlike virtues above all others, the radicality of this pacifist discourse is surprising.[33]

9.2.4 Abbé de Saint-Pierre's Proposal for 'Lasting Peace'

From the end of the seventeenth Century, the debate about peace became a favourite topic of discussion and written texts. The possibility of lasting peace between the European States floated in the air. From Leibniz to Bentham, from the Abbé de Saint-Pierre to Kant, European thinkers elaborated proposals and Academies announced literary prizes devoted to peace. Many of those texts awarded by the French Academy in 1767 fell into oblivion. Other texts, like those of Saint-Pierre, acquired International fame.

Abbé de Saint-Pierre's fame in the Europe of the Age of Enlightenment, which earned him the title of disciple of universal peace (although this form of address would correspond more fairly to other authors), was due to the diffusion of his ideas carried out by Rousseau. Thanks to his reputation as a visionary, his proposals were tolerated in court circles where they were seen as the fanciful and impractical dreams of an honest man.[34] His project of reform sought, on the one hand, to introduce changes to the French monarchy and, on the other hand, to establish a European Union that would guarantee peace. The first part of the plan was set forth in the *Discours sur la Polysynodie*, published in 1718, where he proposed to substitute the existing autocracy of Louis XIV with a cabinet system like the one established by the Regent Phillipe d'Orléans but better. This entailed a modified form of monarchy in which the king reigned but did not govern, and in which he was advised by different councils. The novel character of its contribution lay, on the one hand, in the election system of the advisers, which would have to be carried out by way of a count among the members of a "Political Academy" and, on the other hand, in the rotation of the members of the ministries and councils in order to equip them with a general awareness of all public issues. From his proposal, it

[33]VILLAVERDE, M. J., "Europa y el sueño de la paz perpetua", pp. 31–34.

[34]STELLING-MICHAUD, S., *Introductionn to Ecrits sur l´Abbé de Saint- Pierre, en Oeuvres complètes de J.-J. Rousseau* (Paris: Gallimard, 1964), III, p. CXXIV.

could be deduced that the figure of the monarch could be eliminated without major obstacles and substituted with the president of the "Supreme Council". In practice, the political option that he presented was a mixed government that combined both monarchy and republic.

This interior reform plan was accompanied by a reflection on the international society that figured in the *Projet pour rendre la paix perpétuelle en Europe*, of 1713, and in two other texts of 1717 and 1733. In these he invited Christian sovereigns to sign an agreement to establish a European Society whose senate or regime, endowed with legislative and judicial powers, would resolve conflicts between powers by way of an arbitration tribunal or, whenever necessary, with an army elected by all members which would defend European borders. The presidency would be alternated and independent from the power of each State. The Federation would be made up of all European states with the exception of the Ottoman Empire with whom it would sign a peace agreement similar to the one with the people of North Africa.

In contrast to Sully's plan, to which he made reference, Saint-Pierre considered the Union to be an alliance between powers that respected the international *status quo*. He did not propose changes to the European political map nor to the States themselves. He accepted the political reality and trusted that the good sense of European monarchs would lead them to join in a plan that, he believed, could only yield benefits. Unlike the French moralists (such as La Bruyère or Fénelon), and even philosophers (like Voltaire in *Candide* or the *Dictionnaire philosophique*), who condemned war from a strictly moral point of view, Saint-Pierre operated at an eminently rational level. As Rousseau says, this was not a sensitive soul that burst forth with love for humanity, but rather a cold and methodical one whose sole passion was reason. And if he did preach humanitarianism (he was the one who created the term "charity"), it was because his reason demanded it. He was one of the first and most notable representatives of the Enlightenment; tolerant, a pacifist, and a believer in the progress of universal reason. Despite this, his peace project, like Sully's, lacked the cosmopolitan approach that inspired Crucé and later Kant.[35]

9.2.5 Rousseau's Pessimism and Peace as Unrealisable Utopia

It was an irony of fate that Rousseau was to become Saint-Pierre's interpreter and spokesperson because nobody was further away from the latter's thinking.[36] The

[35]VILLAVERDE, M. J., "Europa y el sueño de la paz perpetua", p. 36.

[36]He received the author's complete works and manuscripts in storage to summarise them and make them readeable. He extracted the *Polysynodie* and the *Projet de paiz perpétuelle,* wrote his *Jugements* on both texts and later renounced the project. The *Polysynodie* and the two *Jugements* remained unpublished until the posthumous edition by Moltou and Du Peyrou, published in Geneva in 1782. Only the *Extrait du Projet de paix perpétuelle* was published at the time, acquiring great popularity. Rousseau felt uncomfortable seeing himself as the spokesman of a thinker whose ideas

Extrait du projet de paiz perpétuelle is a critical summary in which the ideas of both authors interweave and overlap each other to such an extent that it becomes an openly anti-monarchic pamphlet that betrays Saint-Pierre's spirit. Rousseau introduced two fundamental changes to Saint-Pierre's text. On the one hand, he substituted the idea of a "union of powers" for a "federation of sovereign peoples", founding international peace not on the good will of monarchs, but instead on the consent of the people consistent with his republican ideal. On the other hand, he endowed the idea of Europe with more content. While Saint-Pierre proposed an agreement of a political nature, which would lead to a deactivation of conflicts in the name of reason, Rousseau stressed the common features that have invested Europe with cultural identity and have contributed to the forging of a unit and a historical and moral reality; particularly the Roman heritage and Christian tradition.[37] With this, he transcended the economic and geopolitical aspects that shaped the Abbé's vision and invested Europe with a spiritual significance: a conglomerate of peoples united by links such as law, religion, customs, and so on.

However, what really separates the Genevan from the Abbé is something much deeper, linked to the very structure of his thinking: his distrust of reason, his rejection of progress, and his disbelief in human beings' capacity to improve their environment. His radical historical pessimism is what leads him to connect with the Hobbesian paradigm of the inevitability of war. This would distance him inexorably from Saint-Pierre and other Enlightenment thinkers. For Rousseau, the *state of war* that threatens states is irrefutable proof of humankind's failure to attain its moral development. Based on the idea of the natural goodness of human beings that he takes from Seneca; or rather, with the amorality and unsociability of the first individuals who only maintain sporadic contact with each other, Rousseau offloads onto society all the responsibility for the apparition of evil that comes along to mar the pleasant existence of self-sufficient and happy beings.[38] But his argument is perverse as it eludes blaming human beings themselves and accuses society of being the detonator of vices that turn into a source of conflict such as self-love,

---

he perceived as not only unfeasible but in some cases even reprehensible. Although he had reserved the right to indict Saint-Pierre's project and to express his own opinions, in the end, what he most feraed ended up becoming reality; namely, that his name and the Abbé's became united forever to the extent that the ideas of the Frenchman were attributed to him. Both authors were praised together as philosophers of peace, and Rousseau's thought spread in eighteenth century Germany intertwined with Saint-Pierre's philosophy. See CASSIER, E., *Rousseau, Kant, Goethe: Two essays* (Princeton: Princeton University Press, 1945). [Quoted from: E. Cassier, *Rousseau, Kant, Goethe: filosofía y cultura en la Europa del Siglo de las Luces*, trans. y ed. R. R. Aramayo (Madrid: FCE, 2007), pp. 23–25].

[37]"*Extrait du Projet de paix perpétuelle de Monsieur l´Abbé de Saint-Pierre*", in *Oeuvres complètes de J-J. Rousseau*, cit., v. III, p. 567.

[38]In effect, the emergence of the first groups of human beings, as a consequence perhaps of the worsening of the climatological conditions, is the drive that puts into motion the "*perfectability*"; that latent quality of man that, when living in society and transforming his instinct of self-preservation or self-love into pride, becomes the seed of his own ruin (CASSIRER, E., *Rousseau, Kant, Goethe*, p. 28).

pride, the desire to distinguish oneself and prevail over others, and vanity. From this moment on, the conditions for the emergence of private property and inequality are provided, and the outbreak of war of all against all is merely a question of time. Having reached this point there is no turning back.[39]

The portrait that Rousseau paints at the end of the *Discours sur l´origine et les fondemens de l´inégalité parmi les hommes* is similar to that presented by Hobbes in chapter XVII of *Leviathan.* But with a difference; Hobbes, at least, proposes an alternative for establishing peace which is the surrender of freedom for the sake of security. Rousseau, however, lacks a proposal.[40] The only thing he can offer in *Le contrat social* (which is a continuation of the *Discours*) is not developed beyond a mere wish: if only things had happened differently! Essentially, his writing provides an account of how the social pact should have been produced so that, rather than engendering a political society which brought about chaos and conflict, it would bring about a harmonious and peaceful society. But these are mere lamentations. History has led to war within states and between states and international relations are dominated by the law of the strongest.[41]

From his perspective, armed conflicts between European powers are *inherent* to the political system and rooted in the policy of expansion and the conquest of sovereigns driven by spurious interests. The only solution possible would be the subversion of the established political order and the substitution of a Europe of monarchs with a Europe of the people, a collection of republics in which the citizens themselves would govern. But such thinking is mere *desiderátum.* Rousseau rejects this possibility as it would bring even worse evils: "Even if all the advantages of the plan were indisputable, who with any common sense would dare to abolish ancient customs, to change old maxims and to shape the State in a different way from that which it has taken during 1000 and 300 years?" [42] His argumentation leads down an alley with no exit. Nor is there is a solution to the wars between states or to those within them. The only option that he envisages, beyond the eventual treatises, is a union of small nations to defend themselves against large ones; a defensive alliance of the European Republics that had not yet succumbed to the unstoppable process of the degeneration of the monarchies. But such an alliance would not aspire to put an end to war, it was simply proposed as a means of surviving amidst conflict.[43]

If Rousseau renounced his project to elaborate the second part of his book on *Institutions politiques* that would complete his *Contrat social,* it was because he had no alternative to offer except, perhaps, to the peoples who, like the Corsicans or the Poles, had managed to survive ruling moral degradation. But its future was also

[39]See "*Discours sur l´origine et les fondemens de l´inégalité parmi les hommes*" en *Oeuvres complètes*, v. II, p. 507.

[40]CASSIER, E., *Rousseau, Kant, Goethe*, p. 31.

[41]See FERNÁNDEZ, J. F., *Hobbes y Rousseau: entre la autocracia y la democracia* (México: FCE, 1988), pp. 97–98.

[42]"*Jugement sur la Polysynodie*", in *Oeuvres complètes*, p. 638.

[43]FERNÁNDEZ, J. F., *Hobbes y Rousseau*, p. 101.

uncertain. Only the withdrawal into oneself, autarchy, and the absence of relations with the outside could slow down the natural tendency towards degeneration. Herein lies his opposition to commerce. While all authors who have reflected on the question of universal peace (from Crucé to Bentham) regard commerce as a factor of peace, and even as one of the necessary conditions for universal peace, Rousseau defines it as the *Projet de Constitution pour la Corse*; as a "danger to man, a poison for the peoples and a cause of conflict for nations". Whereas the majority considered it to be a unifying element between peoples, by promoting peaceful exchanges and mutual awareness, Rousseau denigrated it as an opening-up factor. Because in this Europe, where the law of the strongest reigned, the only thing that mattered was to close oneself off from the outside. This historical pessimism leads to extremely serious consequences. In particular, it completely negates any undertaking of peace plans or actions that could put an end to war.[44]

Rousseau's theses lead to a constant state of being on the defensive, of being suspicious of the outside, to a fanatical defence of one's own values. This is the approach that informs his two constitution projects for Corsica and Poland, in which a strong sense of fear concerning anything that could lead to uniformity of customs – which he considers to be the first step on the way to annexation – can be perceived: to be alert in order to avoid the propagation of mellifluous foreign habits; to strengthen *national identity* by way of traditions that cannot be uprooted; to erect insurmountable barriers like those built by the great legislators of Antiquity (Numa, Solon and Moses) to protect the people from other nations and to stop them from mixing with them; and above all, to make *love of the nation* the most supreme passion. These are the recipes that Rousseau offers to the Corsicans and Poles in the two texts, his last political writings, where he unveils a whole range of measures for combating foreign ideas and customs and fostering national ones. These proposals extend from the promotion of festivals, games, tastes and all kinds of personal activity, to the recuperation of the national costume, the strengthening of the collective memory through the study of the country's history, the commemoration of great historical achievements, the erection of monuments and the awarding of honours to the nations' martyrs.[45]

Through education, the opinions and tastes of children could be channelled to the point that "they see nothing but the nation from the very moment they open their eyes until their death". In order to convert the inhabitants of a country into "patriots by inclination, out of passion and necessity", education can never be left in the hands of foreigners who could inculcate humanitarian and cosmopolitan ideas.[46] Therefore, as he openly stated in *Lettres écrites de la Montagne*, patriotism and humanitarianism are incompatible.[47] Humanitarian philosophies create fair

[44]VILLAVERDE, M. J., *Rousseau y el pensamiento de las Luces* (Madrid: Tecnos 1987), pp. 75–81.

[45]*Ibíd.*, p. 85.

[46]"*Considérations sur le gouvernement de Pologne*", in *Oeuvres complètes . . .*, v. III, p. 966.

[47]"*Lettres écrites de la Montagne*", in *Ibíd.*, p. 706.

and peace-loving men who may be beneficial for the human race, but are highly prejudicial for nations in the sense of dissolving political ties. Given this, when considering education, it is necessary to choose, as he says in *Emile,* between the formation of men or citizens. And on this dilemma, Rousseau has no doubts. But educating individuals to be citizens involves promoting the patriotic spirit that "is an exclusive spirit that leads us to see almost anyone who is not one of our citizens as foreign or an enemy".[48] The patriots' approach is such that, as he writes in *Emile*, they are hard on foreigners who are merely men or, in other words, unimportant individuals. It is important, he adds, to be good to our neighbours, like the Spartans who, despite being ambitious, greedy and unjust, were unselfish, fair and peaceful with their own.[49]

Rousseau's historical pessimism and his distrust in human reason thereby brings about an exaltation of differences, of that which divides, and of us against them that leads *malgré lui* to division, confrontation, conflict and, in fact, makes the securing of peace impossible. Rousseau does not trust peace because he is convinced that the condition of degradation which the human being has reached has no solution. Because of this, any attempt to regulate international relations or to improve understanding between peoples is in vain. Because deep down, despite the optimistic vision that he presents of original human nature in his second *Discours*, Rousseau does not believe in man. He does not believe in the possibility of moral progress of the individual, however small this may be. Good social institutions, he states in *Emile*, are those that know how to de-naturalise man.[50] In order to do this, it is merely a question of providing him with a new nature and removing his egotism and lack of solidarity to turn him into an ideal citizen of a perfect society.[51]

Rousseau's patriotic vision would, inevitably, end up as a hymn to warlike values. This can be seen in the way he urges states not to fall into decadence, in the way he recommends keeping the people poor and ready for war (efficient remedies against degeneration) and in his proposals to promote a civic religion – even if it be false – to encourage social solidarity and love of the nation. Rousseau did not reach the republican ferocity of Machiavelli nor his war eulogy. His personality, as well as

---

[48] *Correspondance générale de J-J. Rousseau,* Armand Colin, Paris, 1924, IX, p. 265.

[49] *Oeuvres complètes* . . . , v. IV, pp. 248–249.

[50] *Ibíd.*, p. 249.

[51] See VILLAVERDE M. J. "Europa y el sueño de la paz perpetua", p. 47. Hegel also considers the condition of citizenship to be essential, which he endows with *moral character* due to the fact that "the supreme duty of the individuals is to be members of the State" (*Rechtsphilosphie*, 258). Thus, the fundamental duty of the citizen is to obey and to serve the State up to the point of sacrificing one's life for it if this is required (which culminates in dying and killing in war). Rights are *attributed* according to the type of service rendered to the State. Military service is the most excellent manner of rendering service and, therefore, of obtaining maximum rights. War is considered as something *positive* (it consolidates the internal union of the group) and *inevitable*: politics (which fundamentally consists in the defence of the citizen in the face of the foreigner) necessarily implies war against him. Later on, Carl Schmitt took this line of thought even further and established the distinction *friend-enemy* as the key to identity and politics. See BALLESTEROS. J., *Sobre el sentido del derecho*, pp. 45–46.

his historical context, took peace as an object of reflection. It is also true that, in his *Fragments sur la guerre*, he condemned the conflicts that broke out between European powers based on disputes over parcels of power and wealth. The subjects of such States are not interested in whether they come out victorious or not, he says. The war is of no interest to them.[52] And there is no doubt that, in his dedication to Geneva of the second *Discours*, he is in favour of a peaceful state.[53] Nevertheless, his exaltation of patriotism, and the way he equates citizen with patriot, distances him inexorably from pacifism. His confederation of small republics does not intend to be the embryo of a cosmopolitan society composed of citizens of the world, but rather of patriots sheathed in civic pride or nationalist vanity. These small states appear to be armoured against the outside with a reinforced national identity. In contrast to societies open to other cultures that emphasise the rational aspects that unite men, Rousseau advocates closed societies where aspects which separate (religion, language, culture, customs) take on preeminence. It is extremely difficult to base and obtain peace on these premises.

9.3 The Neoconservative Bellicism of the United States: From Rousseau to Schmitt

The North American conception of peace is inscribed in the pessimistic trail of Hobbes and Rousseau, modelled in its latest *NeoCon* perspective by the openly bellicist theses of Hegel and Carl Schmitt as a utopic horizon that must give in to the "*realpolitik*", to the incontestable fact of the inevitability of war. In effect, the exaltation of patriotic values,[54] the fierce vindictive substrate of "an eye for an eye"[55] and an idea of freedom that is synonymous with autonomy and self-sufficiency in the face of a hostile world,[56] has driven American foreign policy to "*go its own way*". That is, to renounce its policy of multilateral alliances and to distrust supranational institutions and bodies in order to guarantee peace, consecrating the principle of "pre-emptive self-defence" (or preventive war) and alluding to its sovereign duties and rights to protect its own territory and citizens

[52]*Oeuvres complètes . . .*, v. III, p. 613.

[53]*Ibíd.*, p. 113.

[54]Regarding this issue see the magnificent film by Wim Wenders, *Land of Plenty* (USA, 2006).

[55]Rifkin indicates that the most emblematic difference between what happens in the Unites States and what happens in the European Union lies, perhaps, in the different attitude towards capital punishment. The death penalty is unanimously rejected in the European Union (Protocol no. 6 of the Protection of Human Rights and Fundamental Freedoms of 1983) whereas the majority of North American society supports it. Even further, France, Russia, Spain and Finland announced that they would not extradite persons suspected of being Al Qaeda terrorists if they were to be subjected to the death penalty (RIFKIN, J., *El sueño europeo*, cit., pp. 364–369).

[56]A view promoted in successful films such as the Oscar-winning *Hostile Territory* (Kathryn Bigelaw, 2009), declared the best film of 2009 by North American critics.

as it considers necessary. This exacerbated sense of sovereignty and patriotism has even led them to affirm that the Constitution of the United States does not allow for the observance of any superior prescription, such as international law, nor for the delegation of their sovereignty to any international organization.[57]

In an excellent essay entitled *Repensar la paz* (*Re-Thinking Peace*), J. Ballesteros identifies the existence of two contrasting bellicisms, which superimpose onto each other, as the fundamental cause that pushes us further away from peace: the *Salafist jihadism*, which justifies violence in order to defend the identity of those nations who feel humiliated (therein lies a significant part of the key to Islamist terrorism)[58] and the bellicism promoted by the *neoconservative ideology* of the United States, rooted in their obsession for national security and for guaranteeing natural resources, especially oil.[59] There are two common elements that the author discovers in both: on the one hand, the division of humanity in terms of Friend-Enemy and, with it, the so-called legality of the double standard (the use of certain violent methods against enemies which would not be legal to use against friends); and on the other hand, the tendency to justify war, the *ius ad bellum*, as *redeeming violence* (moral cleansing and the eradication of evil), without too much consideration for the *ius in bello*, at its own ethical limits.[60] These are, in effect, the two elements that radically separate the European conception of peace from the North American one: peace as premise, means and end (through dialogue, integration and law), versus peace as surrender and submission by means of force, imposition and assimilation.

9.3.1 Neoconservative Ideology

The ideas that fuel *NeoCon* ideology were manifested in Robert Kagan's well-known book *Of Paradise and Power.*[61] Kagan believes that Europe lives in a false

---

[57]The United States' multilateral policy of reciprocal alliances ended with the election of President Bush in 2000 and the eruption of neoconservatives. They thought that North American interests had been neglected and that it was necessary to return to the old foreign policy based on autonomy backed up by military power. This would explain the change in attitude which resulted in their refusal to sign the Kyoto Protocol, the Mine Ban Treaty, the Nuclear Tests Ban Treaty, and their withdrawal from the ABM Treaty, and culminated in their outright refusal to support and submit to the International Criminal Court. Bush made it clear that the United States would use all its military might to establish itself as the indisputable hegemonic power and that he would not allow this supremacy to be challenged by commitments and multilateral treatises or by alliances that subject decision-making to deliberative processes of consensus. (Ibíd., pp. 370–376).

[58]J. Ballesteros carries out an anlysis of this kind of bellicism in Chapter II of *Repensar la paz*, pp. 71–85.

[59]This type of bellicism is treated in Chapter III of the same essay (*Ibíd.*, pp. 87–106).

[60]BALLESTEROS, J., *Repensar la paz*, pp. 87–88.

[61]KAGAN, R., *Of Paradise and Power: America Vs. Europe in the New World Order* (New York: First Vintage Books Editions, 2003, 2004). Translated into Spanish as KAGAN, R., *Poder y debilidad* (Madrid: Taurus, 2003) (see especially pp. 96–125).

postmodern paradise,[62] incapable of defending itself militarily due to its Kantian pacifism and to its refusal to invest seriously in defence. In his view, Europeans hope to extend the formula of Franco-German reconciliation to the whole planet, promoting dialogue, equality in the face of law, and universal submission to the International Criminal Court (which the USA has not ratified). To this pacifist paradise, Kagan opposes the validity of Hobbes' philosophy and suggests that the political priority is to defend life and properties from Islamic terror. Only the United States, aware of their mission and exceptional destiny since its very origins, can preserve the world from terror. As he states: "Americans are idealists but have no experience of promoting ideals successfully without power" and for that reason they have the right to use power unilaterally to deal with possible aggressions coming from the *Axis of Evil*, as the only means to achieve security.[63] In this sense, the role of the United States as global police is good for humanity. Europeans tend to forget what they owe to the United States; more specifically, their unity, the end to the fear of Germany, and the resolving of difficult problems such as Bosnia or Yugoslavia.[64]

9.3.2 Sources of Neoconservative Bellicism: Carl Schmitt and Puritanism

In the above-mentioned essay, J. Ballesteros identifies two of the fundamental sources of *NeoCon* ideology which allow us to understand this bellicist approach: on the one hand, the Hobbesian one, especially the work of Carl Schmitt (the Hobbes of the twentieth century) with his thesis regarding the inevitability of war

---

[62] Kagan refers to an article by M. Cooper, published in *The Observer* (2002) in which he defined three types of State: "pre-modern" ones, always close to chaos such as Somalia and Afganistan; "modern" ones which pursue their national interests with vigour such as China, Brazil and India; and "postmodern" ones (the States that make up the European Union and Japan which have voluntarily given up part of their sovereigny and independence in the name of values such as openess, law and reciprocal security). The United States, according to this writer, has still not decided whether to embrace the postmodern world of interdependency, or to continue with unilateralism and the exercising of military power.

[63] KAGAN, R., *Of Paradise and Power.* [Quoted from: R. Kagan, *Poder y debilidad*, p. 144]. He also states that the superiority of the USA will increase in the future due to its propulation growth which will enable a greater number of armed forces. Its exceptional history and destiny gives it the right to impose its will on the world. The United Nations ought to lose its role as a third party that resolves conflicts as it is a mere creation of the Unites States that has become useless and in which countries which are not clearly allied to the United States - such as France, Russia and China - still have power. The United States, according the author, has still not decided whether to embrace the postmodern world of interdependency, or to continue with unilateralism and the exercising of military power.

[64] Subsequently, Kagan has reiterated his defence for the extention of *ius ad bellum,* even justifying preventive war and has applied these same critical theses to European pacifism in his latest book: *The Return of History and the End of Dreams* (New York: First Vintage Books Editions, 2008).

and the enemy; and on the other hand, Puritanism and the thesis of North American "Manifest Destiny" which highlights the evil and impiety of the enemy and the legality of unlimited war, as long as the cause is holy.[65]

Following Hobbes, Schmitt assumes human evilness and a state of permanent hostility between men, turning war and the enemy into an inevitability. Schmitt, like Hobbes, believes that this anthropological pessimism results from an awareness of original sin, but a very strange one given that it only sees the other's sin and never one's own. For this reason, the disappearance of the external public enemy would imply de-politicization and, therefore, savagery and internal disintegration. What he regards as important, like Hobbes, is the elimination of internal violence within the State, even when this inevitably means waging war against the outside. All that is needed is to defeat internal violence by means of the Leviathan. From this point onwards, Schmitt distances himself from Hobbes and follows Hegel, who not only sees war as inevitable, but also as humanising, given that it contributes to the internal solidarity of the group.[66]

One of the important aspects of Schmitt's thought that inspires the *NeoCon* ideology is a different consideration of war depending on the players. Amongst Europeans, war must be governed by the *ius pubblicum europeum* and has a very "limited" character; on the other hand, war for America, Asia or Africa has an "unlimited" character up to destruction. Schmitt believes that the humanisation of war in Europe was due to the discovery of its ethical excellence compared to other nations after the American Conquest. Europeans would have felt human, thanks to their ethical difference with the rest of humanity, which explains their explicit defence of colonialism. Schmitt (together with Hobbes and Hegel) refuses the concept of humanity as being incompatible with the notion of the enemy. Following this line of thought, *NeoCons* agree with Schmitt (and with Hobbes and Hegel) in denying International Law and refusing any authority above States. For Schmitt, the only third party is the State, whose commitment is to defend its citizens from the enemy. For this reason, he longs for the return of the *ius pubblicum europeum*, which implies a sort of "state of nature" and unlimited war amongst States. War for Schmitt is an unavoidable possibility; it is not the project of politics, but its premise.[67]

The American *NeoCon* theory concerning war adds another category to the bellicist approach of Hobbes and Schmitt (war is necessary and inevitable) that was missing in these authors: its "redeeming" character, tied to the term "manifest destiny" that refers us to the political and economical superiority of the USA as a sign that God desires its expansion. This character is intimately linked to Calvinism,

[65]BALLESTEROS, J., *Repensar la paz*, cit., p. 89. The worse thing about Schmitt (and also in that which is followed by *Neocons*) is that he tends to present his theses as compatible with Christianity when they are in radical opposition to each other. For Schmitt, what is essential is to determine the "enemy of the State", the *hostis,* and more specifically, the enemy within. But Schmitt is wrong to state that the Christian precept of love thy neighbour only forbids hate of a private enemy (*inimicus*) but not a public one (*hostis*).

[66]*Ibíd.*, pp. 92–94.

[67]*Ibíd.*, pp. 95–96.

the religion of the "Founding Fathers" of the United States. As Kagan puts it, the Empire vocation is already found in the origins of the United States and in the fact that its "universalism" derives from its nationalism, from the belief in its exceptional character which has not been weakened.[68]

As Ballesteros rightly points out, the feeling of belonging to a chosen minority and being responsible for wholeheartedly persecuting the non-virtuous (adulterers, drunks, witches, heretics, etc.) as well as non Anglo-Saxons, is essential for Puritanism. According to Calvin, war would be the fervent struggle carried out by a religious society in the name of the Lord of the Armies; and the holier the cause the less restricted the means. When the honour of God is at stake, there is no need to think about humanity. When considering that what is essential in war lies in its cause and in the impiety of the enemy, Calvin comes closer to the Crusades than to the just war. In 1900, Theodor Roosevelt declared that "God has marked the American people as His chosen nation to finally lead in the redemption of the world".[69]

From the standpoint of these messianic-bellicist premises, all that is needed for war is the existence of a just cause, the *ius ad bellum*. The primacy of national security and the world's common good turn the unilateral war against terrorism into an imperative. The existence of such a cause renders the way in which the war is conducted irrelevant, the *ius in bello* (the proportionality) involves a redeeming violence. Therefore, the war against terrorism is perceived as the violence that will end all forms of violence and, for that reason, can and must be unlimited.[70]

As Ballesteros points out, the Neoconservative bellicist conception fully contradicts the better legacy of the United States, summarized in the farewell speeches of two of its greatest presidents, George Washington and Eisenhower. The first of these wrote that "The excessive growth of the military forces is, under any form of government, not at all propitious for freedom, and must be considered as especially hostile to the freedom of the Republic". For his part, Einsenhower lamented the unfortunate novelty of having been constrained to create an immense military organization and an enormous arms industry: "We must fight against the industrial military complex acquiring an unjustified prominence. The potential for a disastrous increase of power already exists and will persist in the future".[71]

Neoconservatism also forgets the legacy of great North American thinkers, defenders of peace and non-violence, such as Thoreau and Luther King. Neither must it be forgotten that the very foundation of the United Nations was, in great measure, an initiative of President Roosevelt, in the same way that his widow had enormous influence in the 1948 Universal Declaration of Human Rights. This represents a whole tradition in favour of peace and support for international authorities that should not be squandered.

---

[68] KAGAN, R., *Of Paradise and Power*, p. 132.

[69] BALLESTEROS, J., *Repensar la paz*, p. 97.

[70] *Ibíd.*, p. 98.

[71] *Ibíd.*, p. 100.

9.4 The Kantian Model: The Possibility of Peace and the European Project

Kant was influenced by Rousseau but, fortunately, without inheriting his pessimism. Although he did not deny the evilness of man, neither did he have a very good opinion of the European monarchs, always ready for war and willing to use their subjects as cannon folder. But he did have confidence in the possibility of introducing improvements both in institutions and morality in order to build a horizon of universal peace. Rousseau would probably smile at such a demonstration of ingenuity. Especially if he had known the extent of Kant's confidence in which even a nation of devils could establish a good political organisation with the mere use of their own reason.[72] And he would refuse to accept that the achievement of universal peace, as Kant suggests, is not only possible, but that a moral duty to try to achieve it exists.[73] Nevertheless, the European dream has been influenced by this source (a mixture of idealism and realism), upon which European thinkers rest today in order to construct their project for transforming history.[74]

At first sight, it seems clear that Kant was convinced that the cosmopolitan ideal was the way towards universal peace. On this basis, he forsaw two consecutive stages in history: (a) the liberation of individuals from the natural state and the constitution of states (b) the end of the natural state in which, once again, the State finds itself in its mutual relations, and the constitution of a universal juridical society.[75] The ideal instrument for moving from one stage to the next must

---

[72]KANT, I., *To Perpetual Peace: A Philosophical Sketch* (Indiana: Hacking Publishing company, 2003). [Quoted from: *La Paz Perpetua*, Madrid, 1985, pp. 83–84].

[73]With the Cosmopolitan World State and universal peace being the goal of Kantian political philosophy, it is surprising that Kant recognised Rousseau (incasable of contemplating peace given his lack of belief in the human being) as his master. Perhaps he may have felt betrayed – in the same way that he was fascinated by the character of Saint-Pierre – despite their idological differences, by Rousseau's boldness in metamorphosing the idea of a Europe of monarchs into a Europe of the people. That would imply establishing the Republican system and placing the freedom of its citizens as the bases of the State, which is to say the faculty of obeying only the laws to which they would have granted their consent. But beyond this, these two authors are diametrically opposed to each other. While Kant believed in respect for individuals' rights, the representative system and separation of powers, Rousseau denied all these principles and did not even recognise the right to meet and associate *(Social Contract*, 1.II, ch. III), nor the right to life *(Social Contract*, 1. II, ch. V.), nor the representative regime *(Social Contract*, 1.III, ch. XV), an abominable regime incompatible with freedom which the collapse of patriotism would have given rise to. According to the Genevan, freedom does not exist if the peoples in question do not ratify their laws. See A. Truyol, "La guerra y la paz en Rousseau y Kant", *Revista de Estudios Políticos* 8 (1994): 57–58.

[74]See HABERMAS, J., "The Kantian Project and the Divided West", *The Divided West*, pp. 113–184.

[75]For Habermas, the Kantian analogy of the natural state applied to relations between States is misleading. Citizens have already taken in an important political trajectory by adhering to the social contract and the legitimating source of power. To advance towards cosmopolitan Law involves a

be International Law, which would be characterised by its provisionality.[76] Thus "[. . .] *all right of peoples and all that can be acquired through war or preserved as external 'mine' and 'yours' among states are merely provisional, and only in a general union of states can this become peremptorily valid (analogous to the union through which a people becomes a state) and can a true condition of peace be attained*".[77] In effect, in the same way that reason demands that men associate with each other in a State in order to end their conflicts, it also requires them to overcome war between nations and to establish a Union of States. It would no longer be enough to build a retaining wall against human perversity, as Hobbes did in *Leviathan*. For Kant, an imperative of reason (a duty) exists that fuels the creation of a cosmopolitan World State, because only within such a framework will it be possible to bring war to an end.[78] This *State of peoples* must be ruled by a cosmopolitan law[79] – already demanded by Crucé – that will guarantee universal hospitality to any human being anywhere on earth.[80]

Two years later, in *On Perpetual Peace*, Kant modifies his position. The construction of a Cosmopolitan State gives way to a more pragmatic and realistic solution: a *federation of sovereign States*. The contradiction between the sovereignty of the States and their subordination to a superior set of rules is resolved by Kant when he affirms that what unites States to each other in a federation is not a juridical obligation, but rather an ethical duty that derives from the agreement between politics and morality.[81] Kant specifies that although this situation of a plurality of States "*is itself a state of war [. . .], this is rationally preferable to the amalgamation of states under one superior power, as this would end in one universal monarchy*".[82]

From this point onwards, Kant establishes the practical conditions for reaching such an end, structuring them as if he was writing an authentic treaty on universal

---

complimentary process, not an analogical one, of a reformulation of the sovereignty and the joining of a cosmopolitan constitution (HABERMAS, J., *The Divided West*, pp. 126–129).

[76]KANT, I., *The Metaphysics of Morals* (Cambridge: Cambridge University Press, 1996). [Quoted from: I. Kant, *La Metafísica de las costumbres* (Madrid: Tecnos, 1989 & 53), p. 181]. Kant understands peace between nations as juridical peace. A conceptual link exists between Law and the attainment of peace, but not in the Hobbesian sense of the submission and obedience of those governed, but rather in the sense of the attainment of rights and freedoms. See HABERMAS, J., *The Divided West*, p. 119.

[77]*Ibíd.*, & 61, p. 190.

[78]*Ibíd.*, 1st part: Theory of Law, sec. 3, conclusion.

[79]The innovative core of this idea consists in the transformation of international law, in terms of the laws of States, into Cosmopolitan Law, in terms of the laws of individuals (HABERMAS, J., *The Divided West*, p.122).

[80]A. Truyol, "A modo de introducción: *La paz perpetua* de Kant en la historia del Derecho de gentes" in *La paz y el ideal cosmopolita de la Ilustración* (Madrid: Tecnos, 1996), p. 27.

[81]For Habermas, this is explained by the presence of a narrow centralist conception of the French Republican model in his mind and its indivisible sovereignty. (HABERMAS, J., *The Divided West*, pp. 124–126).

[82]KANT, I. *To Perpetual Peace*, p. 40.

peace (with six *preliminary* principles or articles for the achievement of peace and three further *definitive* articles for the abolition of war).[83] In addition to the principle of good faith which should guide negotiations (all treatises including clauses which accept the possibility of future war should be invalidated), and two principles characteristic of that time (the exclusion of the purchase of another State and the issuing of public debt securities in relation to foreign policy), the preliminary conditions are: the total and gradual disappearance of all armed forces, the non intervention by means of force in the constitution and governing of other States, and the strict prohibition of dishonourable means in subsequent conflicts, such as war of extermination, that would make future peace impossible.[84]

Once this *preliminary* task had been concluded, historical evolution itself and the need for coexistence and harmony would promote a yearning for peace within the States themselves. In order to attain this, Kant formulates three *definitive principles*:

1. "*The Civil Constitution of Every State Should Be Republican*", with Republic being understood as a political organisation that defends the rights to freedom and equality by means of a separation of powers and a representative government that guarantees innate and imprescriptible rights.[85] On this issue, Kant opposes himself to the thesis of his "master" Rousseau (for whom sovereignty develops by means of direct democracy) and accepts the concept of delegated (representative) democracy.
2. "*The Law of Nations Shall be Founded on a Federation of Free States*". His idea of cosmopolitism does not consist in the elimination of the State but in its integration into a supranational structure.
3. "*Conditions of universal hospitality*" upon which a cosmopolitan law or a law of world citizenship is founded. Namely, that the *foedus pacificum* consists not in a new form of world government, for the aim is not the acquisition of any sort of power, but the establishment of a "federation of states" that would maintain and guarantee the freedom and sovereignty of each of the states, while also enabling a *world citizenship* based upon the very limited and restricted "right of hospitality" or "right to visit", which does not mean the permanent settling of human beings all over the earth (which for Vitoria was the *ius communicationis*). This is, needless to say, a modest cosmopolitan right that sustains a weak form of citizenship compared to the kind that derives from belonging to a State.

---

[83]Here lies one of the most paradoxical aspects of Kant, who considered war reprehensible yet necessary for the progress of humanity. Reconciling his conception of progress linked to war with the pacifist position contained in *La paz perpetua* (Vid. T. Santiago, *Función y crítica de la guerra en la filosofía de I. Kant* (Madrid: Anthropos, 2004), p. 127) is particularly challenging.

[84]KANT, I. *To Perpetual Peace*, pp. 91–96.

[85]KANT, I., *To Perpetual Peace,* First Definitive Article. See also *Teoría y práctica* (Madrid: Tecnos, 1986), p. 27 y ss.

Assuming that the founding of a Universal State and perpetual peace were unattainable ideals (his project should be understood as an "ideal"), and that nations were reluctant to "sacrifice their unlimited wild freedom", that is to say to give up part of their sovereignty and submit to International laws (it is highly likely that nations would hold on to their independence and refuse to constitute themselves in a world State), Kant substitutes the positive idea of a universal Republic with the negative resort of a "*Federation*" of peoples in order to prevent wars.[86]

This "Federation" should not meddle in the internal affairs of States, nor demand the surrender of part of their sovereignty. Its sole mission would be to protect its members from possible attacks. It could be abandoned at any time and would have to be renewed from time to time. This *league of peoples*, or "Völkerbund" – a term which, in German, would officially designate the *Society of Nations* constituted in 1920 – differed from the mere "peace treaty" in the sense that this only put an end to a war while the objective of the "Federation" was to put an end to all wars.[87] According to Kant, the provisional character of this association would not deter the advancement towards the universal Republic, as long as a union could be drawn up around an Enlightened nation – he had in mind the French Republic – that would act as the cohesive core for other States. Nevertheless, until the objective of the universal State was attained, and the natural state persisted among States, the right to war would remain in force and its legality would be recognized in the face of certain cases of aggression or threats to the ruling equilibrium. However, this would have to be exercised in such a way so as not to represent an insurmountable obstacle to the establishment of peace in the future.[88]

After the experience of two World Wars, contemporary European thought finally abandoned the road of *patriotism* backed by Rousseau (the source of intransigence and hostility between nations and the ferment of bellicist nationalisms), and started to take the Kantian way of *cosmopolitism,* based on the intensification of the democratic impulse, respect and mutual recognition, integration (not exclusion and borders), and cooperation and negotiation (rather than imposition and force). And that is because the European dream is to attain peace by overcoming the concepts of sovereignty and citizenship. It aims to establish a world order which is not based on the coercive power of the World State (as representation of the old nation-state) but rather on the universal recognition of the dignity of the human being, as well as universal respect for the human rights that are inherent to it.

---

[86]HABERMAS, J., *The Divided West*, pp. 132–142.

[87]*Ibíd.*, pp. 150–154.

[88]The proposal for a confederation of States versus a single World State enables a difference to be established between International Law – which would constitute the International juridical order through the confederation of States – and Cosmopolitan Law, which would consider the whole of humanity as part of a single political community. In this sense, as Habermas points out, Kant brings a third dimension to the theory of Law, adding to State Law (*ius civitatis)* and to the Law of peoples (*ius gentium*), the Law of the citizens of the world (*ius cosmopoliticum*). Vid. J. Habermas, *El occidente escindido,* cit., pp. 116–134.

The European Union is the first supranational political reality which has dissociated fundamental rights from the sovereignty of the state, and has linked them to a condition of *European citizenship* which is not associated with territoriality, but that is legally binding through the NCHR and the ECHR. The universalisation of Human Rights is the legal articulation of the new European dream, which demands a re-reading of the idea of freedom that can no longer be understood in a negative sense of autonomy and self-sufficiency linked to property (not being dependent on anyone and being prepared to defend oneself against any aggression), and that should, instead, be understood in a positive sense; to be situated in relations of interdependence with others in a way which increases the possibility of satisfying our needs and ambitions more fully and with better quality. As relations increase, so does the degree of freedom that is experienced.[89]

9.5 The Universalisation of the European Peace Model

The basic content of the European model was definitively shaped at the Heads of State and Government Summit of the European Council held in Vienna in 1993 (the year after the Treaty of Maastricht was signed). The text of the final Declaration speaks for itself: "*The end of the division of Europe offers an historic opportunity to consolidate peace and stability in the continent. All our countries are committed to pluralist and parliamentary democracy, the indivisibility and universality of human rights, the rule of law and a common cultural heritage enriched by its diversity. Because of this, Europe can become a vast area of democratic security. This Europe is a source of immense hope which must not be destroyed under any circumstances by territorial ambitions, the resurgence of aggressive nationalism, the perpetuation of spheres of influence, intolerance or totalitarian ideologies*".[90]

The keys to the European model lie in a State of law, democracy and pluralism which together form the basic premise and guarantee for the recognition of universal human rights. Let us consider these elements in their capacity as catalysts for peace.

9.5.1 The State Under the Rule of Law as Instrument of Transformation

The idea of the State under the rule of Law refers us to a fundamental premise that overcomes positivist reductionism. That is, a consideration of Law as a defence of people and their inviolable prerogatives; as something that is above popular

[89]RIFKIN, J., *The European Dream*, pp. 357–359.

[90]Final declaration of the Summit of Heads of State and Government of the European Council held in Vienna (9 de octubre de 1993).

sovereignty. The ethical superiority of the European approach in the struggle against terrorism goes hand in hand with the demand for respect for jurisdictional guarantees, the prohibition of torture, and the abolition of capital punishment. This unconditional respect for every human being, regardless of their behaviour, shows the unanimity with which European legislation opposes torture and the death penalty.[91] The best formula to fight violence and achieve models of political integration lies within the Rule of Law.

In effect, the European model rests upon the desire to move from a world based on the politics of power (in which force is always right) to a community based on the supremacy of this concept of Law. This is directly projected onto the reformulation of the concept of International relations. Europe promotes the international Rule of Law; that is, it does not use military strategy to impose its law on the countries with which it comes into contact, but instead promotes the creation of democratic structures within them, as well as their incorporation into treatises and international bodies of human rights, and the adaptation of their actions to international law. Europe maintains that peace does not require the presence of a global police force that imposes law and order, but rather the transforming power of its politics of peace: to convince people of the advantages of assuming the Rule of Law.[92]

The transforming effect that Europe has had in Turkey is a clear example of this *European model* of doing things. Instead of threatening with military power in order to guarantee their interests, Europe threatens to withdraw its hand of friendship and the prospective of being a member of the EU, unless political reforms take place so that a proper democratic State can be established and human rights can be guaranteed. As Leonard points out, there lies the contrast between the European and North American models. In the face of similar threats (drugs, illegal immigration, networks of international crime), the two responses are diametrically opposed. The United States has sent troops to its neighbouring countries more than 15 times in the last 50 years and the environment has hardly changed (its failure in Colombia contrasts with the success of the EU in Turkey and the Balkans). The European response has been to try to transform their neighbours politically, juridically and institutionally by offering them the possibility of becoming members of the EU and NATO.[93]

Unlike the United States, the prioritising of peace is especially important in the European strategy. The objective of its military force is to build peace rather than to project an image of power. Force might be necessary to build European

---

[91] BALLESTEROS, J., *Repensar la paz*, pp. 114–115.

[92] LEONARD, M., *Why Europe Will Run the 21st Century*, pp. 58–66. This author maintains that Luxembourg, for example, does not obey European and international legislation out of fear of being invaded by North American or European tanks, but out of its own interest in ensuring that international law is maintained. In this sense, he points out that out of the 192 countries of the world, only 12 of them can be considered delinquent states or on the margins of international legality, whereas 180 states respect this legality not out of intimidation by a military power (fear of punishment), but rather in the interest of maintaining legality and peace (p. 66).

[93] *Ibíd.*, pp. 69–71.

values, but it will never be the driving force of European foreign policy. When it deploys its troops it does not do so in order to control other countries, but to eliminate the circumstances that have led those countries to war. The objective of European military action is, above all, to transform the fabric of a society destroyed by war, or, in other words, to extend peace. After the intervention in Kosovo, Europe understood that military intervention had to be included in a wider sequence of commitments oriented towards humanitarian intervention: the use of diplomacy, economic aid, collaboration with the government in question, and a system of sanctions. With this, Europe has managed to impose its doctrine of "preventive commitment" as opposed to the Neoconservative doctrine of "preventive war". The European doctrine of prevention has prevailed because it is based on long term intervention, the military perspective of which is nothing other than the premise that allows for the development of actions in economical, juridical and institutional matters, oriented towards laying the ground for stability and peace in those countries, and thus going far beyond the mere destruction of the source of the threat.[94]

Another of the essential factors in the promotion of peace is the overcoming of ethnocentrism, which is inseparably conjoined with the policy of double standards: of being tolerant and understanding with friends (even when they infringe rights and the law) but implacable with enemies. Peace demands an unconditional assumption of impartiality as a key principle, and of respect for the "third party" (as the figure that incarnates the sense of Law above the conflicting parts). This "third party" is represented today in the international arena by the United Nations (even when defending the need for its reformation) and by the International Criminal Court. It is for this reason that today peace demands that no country can be immune to the decisions and penalties of NATO, nor to the submission of all states to the jurisdiction of the ICC.[95]

9.5.2 *Liberal Democracy as a Way of Excluding Violence*

Democracy constitutes, without doubt, the most appropriate form of political organisation precisely because it offers the most effective means of excluding violence by always placing dialogue at the origin of political decisions.[96] It is on these grounds that Kant considered its establishment as an essential premise for the achievement of perpetual peace.

But if democracy appears to be the most natural way of organising political coexistence, then in order for it to attain its objective of reducing violence to a minimum, it will need to know its own limits which also coincide with the

[94] *Ibíd.*, pp. 79–85.

[95] BALLESTEROS, J., *Repensar la paz*, pp. 121–122.

[96] *Ibíd.*, p. 109.

limits of politics. In this sense, and as Ballesteros rightly reminds us, the opposite of totalitarianism is not democracy, but liberal democracy; namely, that form of democracy that denies Gramsci's affirmation that says that "everything is politics".[97] In other words, overcoming the error of reducing the human being to the condition of citizen (everything is politics) and, consequently, assuming that the will of the majority has universal value, can be decisive. This is why the new concept introduced by Constant, "the freedom of the moderns", which consists in possessing a personal sphere for thinking and acting independently from the power of the State, or even against that power, is of great importance here. In the same way, it is also important to mention the essential character of tolerance and pluralism as two fundamental principles of political liberalism, which are not intrinsically inherent to democracy and which modify it from this perspective.

On these lines, it is worth highlighting that the mere primacy of Law does not eliminate the risks of totalitarianism, which can also appear in democracies. Totalitarian democracy is possible. It emerges wherever the general will aims to become a supreme and universal criterion for justice. This absolutisation of the general will (on the lines of Rousseau's non-liberal democratic thinking) leads to the elimination of dissidents, who end up being considered as criminals. The example of the French Revolutionary Convention is proverbial here. As Ballesteros argues, the democracy that does not recognise its own limits can lead to discrimination and oppression (as is the case with Islamic societies), whereas Law is, structurally speaking, a way of non discrimination and non violence that is more rigorous and complete than democracy itself.[98]

Ballesteros also points out a fundamental question; namely, that the very core of democracy is constituted by freedom of information. Up to the point, that is, that peace and freedom of information are two realities that are mutually required. Freedom of information is a demand for the transparency of power that allows citizens to know the necessary facts that enable them to participate effectively in the political life of their community. It is necessary, then, to establish an incontestable sequence: the more freedom of information, the greater political pluralism, and the more public participation, the better the prospects of peace.[99] And at the same time, there is a connection between transparency of information, democracy, the possibility of internal dissidence, and external peace. As Amnesty International's *International Transparency* report reminded us, transparency of information eradicates the violence of power while secretism engenders abuse. One fact confirms this: the countries in which freedom of information is more greatly infringed are the most violent ones.

---

[97] *Ibíd.*, p. 110.

[98] *Ibíd.*, p. 115. Ballesteros also points out that the very core of this democratic model is constituted by the freedom of information. Up to the point that peace and freedom of information are two realities that are mutually required.

[99] *Ibíd.*, pp. 116–117.

9.5.3 Human Rights as Absolute Premise for Peace

Human rights, as key to the European model, are rooted in one important anthropological principle: the recognition of the "other" (of every "other") as a subject with equal dignity.[100] Dignity, as Ballesteros has argued, is rooted in *being* rather than in *having*.[101] It is anterior to human indicators such as conscience, the capacity for suffering, autonomy, etc., because these qualities affect man's *having* dimension rather than his *being* and, therefore, are not universal features. For this reason, dignity lies in that which all human beings have in common – human nature – and it is connected with the most intrinsic dimension of it: its purpose. Given this, we could say that anyone who is in possession of human nature also possesses ontological dignity and must be consistently treated with maximum dignity, at the same time that this person is also called upon to live their life in accordance with their radical excellence.

Consequently, if dignity is a condition which is essential and inherent to all human beings, it should be recognized in any individual belonging to the *homo sapiens* species, regardless of their state of development or their physical or psychological peculiarities. And, for this reason, if dignity is the ultimate foundation of human rights, then we must conclude that any individual of the human species is a holder of human rights, regardless of their nationality, state of development, and physical or psychological peculiarities.

With regards to peace, it is essential to highlight the validity of human rights above cultural differences. This means recovering the dimension of *humanity*[102] and the need to oppose all relativism with the exception of the analysis of the clash of civilizations, which is underlined by the presupposition of the inevitability of war. Relativism denies the existence of universal values, in particular moral values, and, therefore, the possibility of formulating value judgements endowed with objectivity. When this relativism is projected over cultural diversity, it signals the impossibility of judging not only cultural differences, but specific cultural phenomena such as the different practices, institutions, attitudes and biases that hold currency at a particular time and in a particular culture.

---

[100]The notion of human rights is directly linked to the protection of human beings' essential goods which are demands originating from their *dignity*. "Human dignity" is the only concept which is universally and undisputably recognised as the basis of human rights in all International texts and juridical documents on the subject: such as in the Preamble of the DUHR in its first paragraph: "*Whereas recognition of the inherent dignity and of the equal and inalienable rights of all members of the human family is the foundation of freedom, justice and peace in the world*"; and in the Preamble of the two international agreements of 1966 – that of Civil and Political Rights (ICCPR) and that of Economic, Social and Cultural Rights (ICESCR) – which indicates that: "*The recognition of these rights is based upon the inherent dignity of the human being ...*".

[101]BALLESTEROS, J., *Sobre la fundamentación antropológica de la universalidad de los derechos humanos* (Valencia: Real Academia de Cultura Valenciana, 1999), p. 24.

[102]See D'AGOSTINO, F., "Pluralità delle culture e universalità dei diritti", in *Multiculturalismo e identità* (Milano: Vita e Pensiero, 2002), pp. 45–47.

Evidently, a culture cannot be judged by using another culture as a unit of measurement. To do so would be *ethnocentric*. But neither is a *multiculturalism* acceptable, which limits itself to tolerating and recognising different cultural approaches, proposing a spatio-temporal coexistence between them as stagnant cultural compartments that do not interact with each other. As D'Agostino argues, the mistake of relativism consists precisely in the fact that it presents cultures as "closed universes of experience that refer to nothing other than themselves", when in reality the different cultures represent multiple manifestations of our common humanity. For no culture is self-sufficient, let alone perfect. From the point of view of morality, alongside their positive aspects, all cultures without exception (and that includes Western cultures) have something to reproach themselves for and something that they can modify.[103] No culture constitutes a closed whole, as cultural relativists presuppose; on the contrary, and to a greater or lesser extent, all remain open both to internal changes and external influences. It is true that some cultures are more self-contained than others, probably depending on their own genealogy (even so-called fanatical cultures which are absolutely and hermetically closed to any external influence). Nevertheless, the reality of change is undeniable, even in those cultures that appear to be more self-contained (as is the case of China or India). In this sense, Western culture appears to be especially *open and dynamic*, precisely because its origins were syncretic and because it has always been receptive to influences. And here, at the same time, lies both its eclecticism and its incessant internal crisis.[104]

In denying the existence of universal human values that might serve as a framework for dialogue and understanding amongst human beings belonging to different cultures, radical cultural relativism leads to the isolation, to the absolute separation, of different communities which Touraine calls 'multicommunitarianism'. This implies a fragmentation of society that necessarily generates violence and the risk of social disintegration.[105] This perspective leads to the horizon of new apartheids and to the growing potentiality for new conflicts that push the real achievement of peace further away.

It is therefore also necessary to put an end to ethnocentrism, an approach of those who believe that it is possible to speak of humanity, but who conflate it with their own model of civilization, excluding the other and difference. This model not only presides over North American foreign policy, but is also present in the proposals of prestigious intellectuals like J. Rawls or R. Rorty.[106] In order to

---

[103] D'AGOSTINO, F. *Ibid*, p. 51.

[104] RUBIO-CARRACEDO, J., "Ética intercultural", in *Glosario para una sociedad intercultural*, J. Conill (coord.), pp. 95–97 (Valencia: Bancaja, 2002).

[105] TOURAIN, A., *Pourrons-nous vivre ensemble?*, pp. 201–213 & id., *Igualdad y diversidad*, cit., pp. 50, 53–57 y 81–82. Furthermore, this situation tends to go hand in hand with a tendency to impose homogenisation within the cultural group that does not respect individual autonomy in any way.

[106] See BALLESTEROS, J., "J. Rawls: el conflicto entre pueblos satisfechos y Estados criminales", pp. 471–486.

overcome ethnocentrism, it is necessary to avoid reducing human rights to their individualist and economistic versions as well as to avoid confusing democracy with its substitutes, such as technocracy or the primate of the principle of national security.[107]

I shall conclude, then, by considering the primacy of the Darwinian struggle for survival (seeing the "other" as a competitor) as the inspiring principle of inter-subjective relations, and replacing it with the principle of cooperation (seeing the "other" as a collaborator); thus restoring the primacy of dialogue and confidence as basic premises of the relationship. In order to achieve peace, the European model asserts the universalisation of this principle as the base and foundation of international order (whose ultimate aim is the recognition of universal human rights).

In connection with this, further study is required on the search for *human security* (with the aim of replacing the classical principle of national security), the absolute priority of which is to guarantee the conditions of a worthy life for all human beings without discrimination (drinking water, sufficient food, a healthy environment, sanitary conditions, medicines, etc.). It is essential to put oneself in the place of the "other", with their vulnerability, suffering and misery, to be able to appropriately prioritise development aid and disarmament as the basic requirements of international justice (the arms race is the main cause of underdevelopment in the world). The true way to extend peace is not to continue bombing countries, but to eradicate injustices and fight against underdevelopment.[108]

---

[107] BALLESTEROS, J. *Repensar la paz*, cit., pp. 108–109.

[108] *Ibíd.*, pp. 120–121.

Author Bios

Jesús Ballesteros

Graduate in Law (1965) and Doctor in Law (1972) with extraordinary prizes upon graduation at the University of Valencia, Spain. He is professor of Philosophy of Law and Political Philosophy at University of Valencia, Spain (since 1983) and Director of Doctoral Programme of *Human Rights, Peace and Sustainable Development* (since 1998). He was Director of Department of Philosophy of Law, Moral and Politics at the University of Valencia (1985–1993) and he has been Director of numerous research projects related to Human Rights, the Environment, Genetics and Global Justice, financed by *Generalitat Valenciana* and the Spain Government (since 2001 until nowadays). Member of the Board of Editors of following Reviews: *Anuario de Derechos Humanos, Cuadernos de Bioética, Persona y Derecho, Ragion Pratica, Revista de Derecho Ambiental* and *Teoria del Diritto e dello Stato. Rivista europea de cultura e scienza giuridica*. Member of the International Association of Philosophy of Law and Social Philosophy (IUR) and of Spanish Society of Bioethics (AEBI). He is editor of eight books and author of numerous monographs (8), collective books (47) and scientific articles in specialized national and international magazines (40). Also, he has directed more than 30 research works and 20 doctoral thesis and he has participated as a speaker in many local and international conferences.

Francesco Biondo

He is Tenured Full-time Lecturer of Philosophy of Law at University of Palermo, Italy. Phd on "Diritti umani. Evoluzione, tutela e limiti" (1997–2000). Research Associate at the Balliol College, Oxford (1998–1999 and 2000–2001) with a dissertation published as book with the title "Benessere, Giustizia e diritti umani nel pensiero di Amartya Sen", Giappichelli, Torino 2003; Post-doctoral fellowship at the University Sour Orsola Benincasa, Napoli (2003–2004) with a dissertation on "Disobbedienza civile ≪conservativa≫ e neocostituzionalismo: rilevanza e limiti di un dibattito teorico", published in V. Omaggio (a cura di) *Diritto in trasformazione. Questioni di filosofia giuridica*, Editoriale Scientifica, Napoli 2004,

J. Ballesteros et al. (eds.), *Globalization and Human Rights*, Ius Gentium: Comparative Perspectives on Law and Justice 13, DOI 10.1007/978-94-007-4020-4, pp. 285–339. He has been Visiting Scholar at the Balliol College (Oxford University) (1999 and 2001) and Visiting Research Fellow at the University of Valencia, Department of Law, Moral and Political Philosophy (2010).

Jorge Cardona Llorens

He is currently Professor of Public International Law and International Relations at the University of Valencia, Spain and Member of the Committee on Rights of Children, U.N. Geneva. He got a Special Graduation Award (1980) and a Special PhD Award at the University of Valencia (1984). Teacher of Public International Law at the Universities of Valencia (1980–1984; 1985–1995; and since 2008), Alicante (1984–1985) and Jaume I in Castellón (1995–2008). Director of the International Law Section at the *Bancaja International Centre for Peace and Development* (since 1997). Director (and founder) of the *Bancaja Euro-Mediterranean Courses on International Law* (since 1997). *Holder of the Henri Rolin Chair* at the Universities of Louvain-La-Neuve and Vrije Universiteit Brussel (2009), teaching seminars on ≪Quelle place pour les droits de l'homme et le droit international humanitaire dans les situations d'administration internationale de territoire?≫. *Holder of the Jean Monnet Chair* in Community Law (since 1997). President of the *Mediterranean Institute of European Studies* (since 1999). President of the *Centre for Studies on Law and International Relations* (www.cedri.org) (since 2007). *Visiting Professor* at different Universities and University Institutes in Europe, Latin America and Africa. He has led and participated in more than 20 research projects funded by public and private bodies. Member of the scientific committee of the *Revista Española de Derecho Internacional* (1990–1998). Member of the editorial board of the *Gaceta Jurídica de la Unión Europea*, since 1989. He has authored around 150 publications on international law, in particular on the law of treaties, the law of international responsibility, the law of international organizations, human rights, peacekeeping and international security, among others. He has participated 200 national and international conferences and debates, and he has directed around 50 research works and 10 doctoral thesis.

Encarnación Fernández Ruiz-Gálvez

Associate Professor of Philosophy of Law and Political Philosophy at University of Valencia, Spain; Director of Department of Philosophy of Law, Moral and Politics (since March 2007) and Director of Master *Human Rights, Peace and Sustainable Development* (since September 2009) at University of Valencia, Spain. Her teaching and research focus on human rights and the equality principle. More recently she has also researched different aspects of the Theory of the State and the Philosophy of the International Relations, with special attention to the failed States (characters, reasons, consequences, reconstruction of the State). Author of numerous monographs, collective books and scientific articles in specialized national and international magazines. A select list of her publications are listed below: *Igualdad y Derechos Humanos*, Tecnos, Madrid, 2003, 198 pp.; *¿Estados fallidos o Estados en crisis?*, Comares, Granada, 2009, 183 pp.; *Justicia, Solidaridad, Paz. Estudios en Homenaje* *al Prof. José Mª Rojo Sanz* (co-editor), Valencia, 1995, 871 pp.; and *Biotecnología y posthumanismo* (co-editor), Thomson-Aranzadi, Pamplona, 2007, 503 pp. She has led and participated in many research projects funded by Italy and Spain's governments. Also, she is Member of Scientific Committee of Law Studies Centre *Di-con-per-Donne* of University of Rome Tor Vergata and Member of European Network EuropePolis. She has participated as speaker in numerous national and international conferences and she has directed more than ten research works and doctoral thesis.

Ana-Paz Garibo Peyró

Graduated in Law (1987), PhD in Philosophy of Law (2002), and Master in "Law and Bioethics" (2000). Currently is Tenured Full-time Lecturer of Philosophy of Law and Political Philosophy at University of Valencia, Spain. She was Visiting Lecturer at Managua University (2003) and Tampico University (2004/2005). Also, she was Visiting Researcher at Oxford University (2008/2009) and University of Rome Tor Vergata (2009/2010 and 2010/2011). The focus of her research and teaching is the rights of the child, children in armed conflict, women's rights, Feminism and pacifism in the Islam. She has published two books and several articles related to human rights.

Isabel Trujillo

Professor of Philosophy of Law at the Faculty of Law of University of Palermo. Director of Doctoral School "Supranational and National Law", Law Faculty, University of Palermo. Editorial Board of "Ragion Pratica", "Diritto e questioni pubbliche". PhD in Legal Philosophy, General Theory of Law and Political Philosophy at the University of Rome, "La Sapienza", Italy (1997). Post-doc University of Ferrara (1998–1999). Researcher in University of Rome "Tor Vergata" (1999–2001). Jemolo Fellowship, Nuffield College, Oxford (2007). Academic Visitor Law Faculty, Oxford (2009).

Francesco Viola

Professor of Philosophy of Law at the University of Palermo since 1986 and also appointed to the courses of General Theory of Law, Human Rights and Legal Ethics. He is co-editor of a Series on Legal Philosophy of Giappichelli Publishing in Turin, co-director of "Ragion pratica", a journal on practical philosophy and co-director of "Ars Interpretandi. Annuario di ermeneutica giuridica", a journal on legal hermeneutics. Since 2001 he has been director of the Section "Law and Politics" of the new edition of the Encyclopedia of Philosophy (Bompiani). From 2007 to 2009 he has been director of the PhD Programs of the Faculty of Law on "Supranational and Domestic Law". At the moment he is director of the Department of Studies on Politics, Law and Society. He is also President of the Italian Association for Legal Philosophy.

Ernesto. J. Vidal Gil

Professor of Philosophy of Law and Political Philosophy at the University of Valencia. He was Director of Department of Philosophy of Law, Moral and Politics at the University of Valencia (2001–2007). He is author of many monographs, collective books and scientific articles in specialized national and international magazines. His principal lines of research have been the personalistic thought, the reformulation of a general theory of the Law from the Constitution, the link between human rights and positive duties, the rights of solidarity, the immigrants' rights and, different problems of the biolaw. Also, he is Substitute Magistrate of the Court of Justice of Valencian Community in Spain and Member of Hospital Ethics Committee of University of Valencia, Spain.

Pedro Talavera

Associate Professor of Philosophy of Law and Political Philosophy at the University of Valencia since 2000. He teaches the courses of General Theory of Law, Human Rights and Legal Philosophy. PhD from University *La Sapienza* of Rome (Italy) and JD from University of Valencia (Spain). Visiting professor at Loyola University of Chicago (USA); University of Cambridge (UK); Catholic University of Chile and University *Gabriel René Moreno* of Santa Cruz (Bolivia). Between his representative monographs: *La metodología de Ireneo de Lyon* (Barcelona 1995); *Las uniones homosexuales frente al Derecho* (Madrid 1999); *La unión de hecho y el Derecho* (Granada 2001); *Derecho y Literatura*, (Granada 2006), *Interpretación, integración y argumentación jurídica* (Santa Cruz 2008); *Honor, intimidad e imagen* (Santa Cruz 2011). He has also published about nationalism, identity and multiculturalism; bioethics issues, the objection of conscience and the juridical treatment of immigration.

Index

J. Ballesteros et al. (eds.), *Globalisation and Human Rights*, Ius Gentium: Comparative Perspectives on Law and Justice 13, DOI 10.1007/978-94-007-4020-4,

Printed by Printforce, the Netherlands